Lecture Notes in Civil Engineering

Volume 311

Lecture Notes in Civil Engineering (LNCE) publishes the latest developments in Civil Engineering—quickly, informally and in top quality. Though original research reported in proceedings and post-proceedings represents the core of LNCE, edited volumes of exceptionally high quality and interest may also be considered for publication. Volumes published in LNCE embrace all aspects and subfields of, as well as new challenges in, Civil Engineering. Topics in the series include:

- Construction and Structural Mechanics
- Building Materials
- Concrete, Steel and Timber Structures
- Geotechnical Engineering
- Earthquake Engineering
- Coastal Engineering
- Ocean and Offshore Engineering; Ships and Floating Structures
- Hydraulics, Hydrology and Water Resources Engineering
- Environmental Engineering and Sustainability
- Structural Health and Monitoring
- Surveying and Geographical Information Systems
- Indoor Environments
- Transportation and Traffic
- Risk Analysis
- Safety and Security

To submit a proposal or request further information, please contact the appropriate Springer Editor:

- Pierpaolo Riva at pierpaolo.riva@springer.com (Europe and Americas);
- Swati Meherishi at swati.meherishi@springer.com (Asia—except China, Australia, and New Zealand);
- Wayne Hu at wayne.hu@springer.com (China).

All books in the series now indexed by Scopus and EI Compendex database!

Magd Abdel Wahab
Editor

Proceedings of the 5th International Conference on Numerical Modelling in Engineering

Volume 1: Numerical Modelling in Civil Engineering, NME 2022, 23–24 August, Ghent University, Belgium

Springer

Editor
Magd Abdel Wahab
Soete Laboratory
Faculty of Engineering and Architecture
Ghent University
Zwijnaarde, Belgium

ISSN 2366-2557 ISSN 2366-2565 (electronic)
Lecture Notes in Civil Engineering
ISBN 978-981-19-8428-0 ISBN 978-981-19-8429-7 (eBook)
https://doi.org/10.1007/978-981-19-8429-7

This Springer imprint is published by the registered company Springer Nature Singapore Pte Ltd.
The registered company address is: 152 Beach Road, #21-01/04 Gateway East, Singapore 189721, Singapore

Preface

This volume contains the proceedings of the 5th International Conference on Numerical Modelling in Engineering: Volume 1 Numerical Modelling in Civil Engineering. Numerical Modelling in Engineering NME 2022 is the 5th NME conference and is held Online via MS Teams, during the period 23–24 August 2022. Previous NME conferences were celebrated in Ghent, Belgium (2018), Beijing, China (2019) and Ghent, Belgium (2020–2021).

The overall objective of the conference is to bring together international scientists and engineers in academia and industry in fields related to advanced numerical techniques, such as FEM, BEM, IGA, etc., and their applications to a wide range of engineering disciplines. The conference covers industrial engineering applications of numerical simulations to Civil Engineering, Aerospace Engineering, Materials Engineering, Mechanical Engineering, Biomedical Engineering, etc. The presentations of NME 2022 are divided into two main sessions, namely (1) Civil Engineering and (2) Mechanical and Materials Engineering. This volume is concerned with the applications to Civil Engineering.

The organising committee is grateful to keynote speaker, Prof. Timon Rabczuk, Bauhaus Universität Weimar, Chair of Computational Mechanics, Germany, for his very interesting keynote speech entitled 'Machine Learning Based Solutions of Partial Differential Equations'.

Special thanks go to members of the Scientific Committee of NME 2022 for reviewing the articles published in this volume and for judging their scientific merits. Based on the comments of reviewers and the scientific merits of the submitted manuscripts, the articles were accepted for publication in the conference proceedings and for presentation at the conference venue. The accepted papers are of a very high scientific quality and contribute to the advancement of knowledge in all research topics relevant to NME conference.

Finally, the organising committee would like to thank all authors, who have contributed to this volume and to those who have presented their research work at the conference in MS Teams.

Zwijnaarde, Belgium

Prof. Magd Abdel Wahab
Chairman of NME 2022

Organising Committee

Chairman

Prof. dr. ir. Magd Abdel Wahab, Laboratory Soete, Ghent University, Belgium

International Scientific Committee

Prof. D. Ribeiro, School of Engineering, Polytechnic of Porto (ISEP-IPP), Portugal
Prof. J. Santos, University of Madeira, Portugal
Prof. J. Toribio, University of Salamanca, Spain
Prof. B. B. Zhang, Glasgow Caledonian University, UK
Prof. V. Silberschmidt, Loughborough University, UK
Prof. T. Rabczuk, Bauhaus University Weimar, Germany
Prof. L. Vanegas Useche, Universidad Tecnológica de Pereira, Colombia
Prof. N. S. Mahjoub, Institut Préparatoire aux Etudes d'Ingénieurs de Monastir, Tunisia
Prof. A. Cheknane, Amar Telidji University of Laghouat, Algeria
Prof. E. N. Farsangi, Kerman Graduate University of Advanced Technology (KGUT), Iran
Prof. N. A. Noda, Kyushu Institute of Technology, Japan
Prof. K. Oda, Oita University, Japan
Prof. S. Abdullah, Universiti Kebangsaan Malaysia, Malaysia
Prof. C. Zhou, Nanjing University of Aeronautics and Astronautics, China
Prof. B. Bhusan Das, National Institute of Technology Karnataka, India
Prof. R. V. Prakash, Indian Institute of Technology, India
Prof. H. N. Xuan, Hutech University, Vietnam
Prof. Giuseppe Carbone, University of Calabria, Italy
Prof. Fadi HAGE CHEHADE, Lebanese University, Lebanon
Prof. Sohail Nadeem, Quaid-i-Azam University, Pakistan

Dr. A. San-Blas, Miguel Hernández University of Elche, Spain
Dr. G. Minafo, University of Palermo, Italy
Dr. A. Caggiano, Technische Universität Darmstadt, Germany
Dr. S .Khatir, Ghent University, Belgium
Dr. T. Yue, Ghent University, Belgium
Dr. A. Rudawska, Lublin University of Technology, Poland
Dr. L. V. Tran, Sejong University, South Korea
Dr. X. Zhuang, Leibniz Unversität Hannover, Germany
Dr. I. Hilmy, International Islamic University Malaysia, Malaysia
Dr. C. Wang, Liaocheng University, China
Dr. M. Mirrashid, Semnan University, Iran
Prof. A. G. Correia, University of Minho, Portugal
Dr. M. Wang, Los Alamos National Laboratory, USA
Dr. Filippo Genco, Adolfo Ibáez University, USA
Dr. Denis Benasciutti, University of Ferrara, Italy
Dr. Y. L. Zhou to Xi'an Jiaotong University, China

Contents

About the Editor

Prof. Magd Abdel Wahab is a Full Professor of Applied Mechanics in the Faculty of Engineering and Architecture at Ghent University, Belgium. He received his B.Sc., 1988, in Civil Engineering and his M.Sc., 1991, in Structural Mechanics, both from Cairo University. Prof. Wahab completed his Ph.D. in Fracture Mechanics in 1995 at KU Leuven, Belgium. He was awarded the degree of Doctor of Science from the University of Surrey in 2008. He has published more than 600 scientific papers in solid mechanics and dynamics of structures and edited more than 30 books and proceedings. His research interests include fracture mechanics, damage mechanics, fatigue of materials, durability, and dynamics and vibration of structures.

Nonconforming Spectral Element Method for Oseen Equations and Navier-Stokes Equations

N. Kishore Kumar and Subhashree Mohapatra

Abstract In this paper, we discuss the performance of a least-squares-based spectral element solver for Oseen equations on two-dimensional curvilinear domains and three-dimensional Navier-Stokes equations. Both equations are solved in primitive form without any first-order reformulation. The spectral approximation is nonconforming, and the same order spectral element functions are used for both velocity and pressure variables. A suitable preconditioner has been proposed using ADN theory in order to control the condition number of the system. Numerical results are obtained using the preconditioned conjugate gradient method. Numerical results show that the method is exponentially accurate in both velocity and pressure variables. Mass conservation property of the used solver has been displayed.

1 Introduction

The Oseen and Navier-Stokes equations arise in many engineering applications. These equations have been widely studied using various numerical methods like finite difference methods, mixed finite element methods, least-squares methods, spectral methods and different nonconforming methods in the literature [5, 19, 26]. Our main focus is on least-squares-based numerical schemes.

Numerical methods based on the least-squares approach for elliptic differential equations, in elliptic Stokes and Navier-Stokes equations have been discussed in [4, 11, 18]. Least-squares methods offer an alternative approach to standard mixed formulations and many other computational advantages such as resulting systems being always symmetric and positive definite, allowance of equal order interpolation, etc. [2, 8, 9, 11, 22–24].

N. Kishore Kumar
BITS-Pilani Hyderabad Campus, Hyderabad, India
e-mail: naraparaju@hyderabad.bits-pilani.ac.in

S. Mohapatra (✉)
IIIT, Delhi, India
e-mail: subhashree@iiitd.ac.in

M. Abdel Wahab (ed.), *Proceedings of the 5th International Conference on Numerical Modelling in Engineering*, Lecture Notes in Civil Engineering 311,
https://doi.org/10.1007/978-981-19-8429-7_1

Least-squares-based finite element methods for different first-order reformulations have been discussed in [3, 14, 15, 17, 20, 27]. A stabilized method for Oseen equations has been proposed in [12]. A local discontinuous Galerkin method and a hybridized discontinuous method have been proposed in [6, 7] for Oseen equations while transforming into a first-order system.

In this article, we investigate an exponentially accurate spectral element solver for Oseen equations on two-dimensional curvilinear domains and three-dimensional Navier-Stokes equations. This method is nonconforming, and the numerical formulation is least-squares. The equations are considered in primitive form without converting them into a first-order system. The minimizing functional in the least-squares formulation includes the residues in the differential equations, residues in the boundary conditions and jumps in velocity components, pressure variable and first-order derivatives of the velocity components along the element interfaces in the suitable Sobolev norms. The numerical solution is obtained at Gauss-Legendre-Lobatto quadrature points using the preconditioned conjugate gradient method.

Notations and stability estimates are introduced in Sect. 2. We propose the numerical scheme and error estimates in Sect. 3. Section 4 displays the numerical results. We conclude with Sect. 5.

2 Preliminaries and Stability Estimate

2.1 Notations

Let $\Omega \in \mathbb{R}^d$, $d = 2, 3$ be an open bounded set of class C^2 with the boundary $\partial\Omega$. $H^m(\Omega)$ denotes the Sobolev space of functions with square integrable derivatives of integer order less than or equal to m on Ω equipped with the norm

$$\|u\|^2_{H^m(\Omega)} = \sum_{|\alpha|\le m} \|D^\alpha u\|^2_{L^2(\Omega)} \,.$$

Further, let $E = (-1, 1)^{d-1}$. Then we define fractional norms $(0 < \sigma < 1)$ by

$$\|w\|^2_{\sigma,\Gamma_l} = \|w\|^2_{0,E} + \int_E \int_E \frac{|w(\xi) - w(\xi')|^2}{|\xi - \xi'|^{1+2\sigma}} d\xi d\xi'. \tag{1}$$

Here, $w = u$, v or p. Moreover,

$$\|w\|^2_{1+\sigma,\Gamma_l} = \|w\|^2_{0,E} + \left\| \frac{\partial w}{\partial \xi} \right\|^2_{\sigma,E} + \left\| \frac{\partial w}{\partial \eta} \right\|^2_{\sigma,E} .$$

We shall denote the vectors by bold letters. For example, if $\Omega \subset \mathbb{R}^2$, $\boldsymbol{u} = (u_1, u_2)^T$, $\boldsymbol{H}^k(\Omega) = H^k(\Omega) \times H^k(\Omega)$, $||\boldsymbol{u}||^2_{k,\Omega} = ||u_1||^2_{k,\Omega} + ||u_2||^2_{k,\Omega}$, etc. Spaces and norms are defined in a similar way for three-dimensional cases. Let us denote a point $\boldsymbol{x}$ in the domain Ω by $\boldsymbol{x} = (x_1, x_2) \in \Omega \subset \mathbb{R}^2$ and $\boldsymbol{x} = (x_1, x_2, x_3) \in \Omega \subset \mathbb{R}^3$.

2.2 *Oseen Equations*

Here, we consider Oseen equations with Dirichlet boundary conditions

$$-\frac{1}{Re}\Delta \boldsymbol{u} + (\boldsymbol{b} \cdot \nabla \boldsymbol{u}) + \nabla p = \boldsymbol{f} \quad \text{in} \quad \Omega, \tag{2}$$

$$-\nabla \cdot \boldsymbol{u} = h \quad \text{in} \quad \Omega, \tag{3}$$

$$\boldsymbol{u} = \boldsymbol{g} \quad \text{on} \quad \partial\Omega. \tag{4}$$

Here, $\boldsymbol{u}$ is the velocity field, p is the pressure, $\boldsymbol{f} \in \boldsymbol{L}^2(\Omega), \boldsymbol{g} \in \boldsymbol{H}^{\frac{1}{2}}(\partial\Omega)$, $\boldsymbol{b} = (b_1, ..., b_d)$ is a given C^1 vector function and 'Re' is the Reynolds number.

Let $\mathcal{L}(\boldsymbol{u}, p)$ and $\mathcal{D}\boldsymbol{u}$ be the differential operators for the momentum equations and the continuity equation, respectively. So,

$$\mathcal{L}(\boldsymbol{u}, p) = -\frac{1}{Re}\Delta \boldsymbol{u} + (\boldsymbol{b} \cdot \nabla)\boldsymbol{u} + \nabla p,$$
$$\mathcal{D}(\boldsymbol{u}) = -\nabla \cdot \boldsymbol{u}.$$

Here, we state a regularity estimate [26] that is derived using ADN theory [1].

Let $\Omega \subset \mathbb{R}^d$ be bounded with $\partial\Omega$ of class C^2. For $(\boldsymbol{u}, p) \in \boldsymbol{H}^2(\Omega) \times H^1(\Omega)$ being solutions of the Oseen problem (2)–(4), if $\boldsymbol{f} \in \boldsymbol{L}^2(\Omega)$, $h \in H^1(\Omega)$, $g \in \boldsymbol{H}^{3/2}(\Gamma)$, then $(\boldsymbol{u}, p)$ satisfies

$$\|\boldsymbol{u}\|_{\boldsymbol{H}^2(\Omega)} + \|p\|_{H^1(\Omega\backslash\mathbb{R})} \leq C_0 \left(\|\boldsymbol{f}\|_{\boldsymbol{L}^2(\Omega)} + \|h\|_{H^1(\Omega)} + ||\boldsymbol{g}||_{\boldsymbol{H}^{3/2}(\Gamma)}\right). \tag{5}$$

2.3 *Discretization and Stability Estimate*

In order to keep the notations and presentation simple, we stick to a domain $\Omega \subset \mathbb{R}^2$, whereas all theoretical estimates proposed here hold in three-dimensional domains too.

We divide domain Ω into L number of rectangles, $K_l, l = 1, \cdots L$ and define spectral element functions as a tensor product of polynomials. Let $K = (-1, 1) \times (-1, 1)$ and $\exists$ an analytic map M_l from K_l to K with an analytic inverse. A set of spectral element functions are defined on these elements which are a sum of tensor products of polynomials. The map M_l is of the form $\hat{\boldsymbol{x}} = M_l(\boldsymbol{x})$, where $\hat{\boldsymbol{x}} = (\xi, \eta)$.

Define the spectral element functions $\hat{\boldsymbol{u}}$ and $\hat{p}$ on K by

$$\hat{\boldsymbol{u}}(\xi,\eta)=\sum_{i=0}^{W}\sum_{j=0}^{W}\mathbf{a}_{i,j}\,\xi^i\eta^j,\quad \hat{p}(\xi,\eta)=\sum_{i=0}^{W}\sum_{j=0}^{W}b_{i,j}\xi^i\eta^j.$$

Then $\boldsymbol{u}_l$ and p_l on K_l are given by

$$\boldsymbol{u}_l(x_1,x_2)=\hat{\boldsymbol{u}}(M_l^{-1})\ \ and\ \ p_l(x_1,x_2)=\hat{p}(M_l^{-1}).$$

Let $\boldsymbol{\Pi}^{L,W}=\left\{\{\boldsymbol{u}_l\}_{1\le l\le L},\{p_l\}_{1\le l\le L}\right\}$ be the space of spectral element functions consisting of the above tensor products of polynomials.

Let Γ_l be a common to adjacent elements K_l and K_m. Assume that the edge Γ_l is the image of $\eta=1$ under the map M_l which maps K to K_l and also the image of $\eta=-1$ under the map M_m which maps K to K_m.

Let $I=(-1,1)$. Define the jump along the interelement boundaries as

$$\begin{aligned}
\|[\boldsymbol{u}]\|_{0,\Gamma_l}^2&=\left\|\hat{\boldsymbol{u}}_m(\xi,-1)-\hat{\boldsymbol{u}}_l(\xi,1)\right\|_{0,I}^2,\\
\left\|[\boldsymbol{u}_{x_k}]\right\|_{\frac{1}{2},\Gamma_l}^2&=\left\|(\hat{\boldsymbol{u}}_m)_{x_k}(\xi,-1)-(\hat{\boldsymbol{u}}_l)_{x_k}(\xi,1)\right\|_{\frac{1}{2},I}^2,\\
\|[p]\|_{\frac{1}{2},\Gamma_l}^2&=\left\|\hat{p}_m(\xi,-1)-\hat{p}_l(\xi,1)\right\|_{\frac{1}{2},I}^2.
\end{aligned}$$

For $\boldsymbol{u},p\in\boldsymbol{\Pi}^{L,W}$, define two quadratic forms

$$\begin{aligned}
\mathcal{V}^{L,W}(\boldsymbol{u},p)&=\sum_{l=1}^{L}||\mathcal{L}(\boldsymbol{u}_l,p_l)||_{0,K_l}^2+\sum_{l=1}^{L}||\mathcal{D}\boldsymbol{u}_l||_{1,K_l}^2\\
&+\sum_{\Gamma_l\subseteq\bar{\Omega}\setminus\partial\Omega}\left(|||[\boldsymbol{u}]|||_{0,\Gamma_l}^2+\sum_{k=1}^{2}|||[\boldsymbol{u}_{x_k}]|||_{\frac{1}{2},\Gamma_l}^2+|||[p]|||_{\frac{1}{2},\Gamma_l}^2\right)+\sum_{\Gamma_l\subseteq K_l\cap\partial\Omega}||\boldsymbol{u}_l||_{\frac{3}{2},\Gamma_l}^2
\end{aligned}\tag{6}$$

and

$$\mathcal{U}^{L,W}(\boldsymbol{u},p)=\sum_{l=1}^{L}||\boldsymbol{u}_l||_{2,K}^2+\sum_{l=1}^{L}||p_l||_{1,K}^2.\tag{7}$$

We have the following theorem from [21].

Theorem 2.1 *For W large enough, there exists a constant $C>0$ such that the estimate*

$$\mathcal{U}^{L,W}(\boldsymbol{u},p)\le C(\ln W)^2\mathcal{V}^{L,W}(\boldsymbol{u},p)\tag{8}$$

holds.

3 Numerical Scheme, Error Estimates and Computational Aspects

3.1 Numerical Scheme

In this section, we define the numerical scheme which is the following least-squares functional, which needs to be minimized,

$$\mathcal{R}^{L,W}(\boldsymbol{u},p)=\sum_{l=1}^{L}\|\boldsymbol{L}\boldsymbol{u}_l-\boldsymbol{F}_l\|_{0,K}^2+\sum_{l=1}^{L}||\boldsymbol{D}\boldsymbol{u}_l-h_l||_{1,K}^2$$
$$+\sum_{\Gamma_l\subseteq\bar{\Omega}\backslash\partial\Omega}\left(||[\boldsymbol{u}]||_{0,\Gamma_l}^2+\sum_{k=1}^{2}||[\boldsymbol{u}_{x_k}]||_{\frac{1}{2},\Gamma_l}^2+||[p]||_{\frac{1}{2},\Gamma_l}^2\right)+\sum_{\Gamma_l\subseteq K_l\cap\partial\Omega}||\boldsymbol{u}_l-\boldsymbol{g}_l||_{\frac{3}{2},\Gamma_l}^2. \tag{9}$$

Here, $\boldsymbol{L}\boldsymbol{u}_l$, $\boldsymbol{D}\boldsymbol{u}_l$ are differential operators for the momentum and continuity equations in ξ and η variables, respectively. $\boldsymbol{F}_l$, h_l and $\boldsymbol{g}_l$ are the source terms and boundary data respectively in ξ and η variables.

Next, we state the error estimate for the proposed scheme which is exponentially small in W [16, 21].

Theorem 3.1 *Let (z,q) minimize $\mathcal{R}^{L,W}(\boldsymbol{u},p)$. Then for W large enough, there exist constants C and b (being independent of W) such that the estimate*

$$\sum_{l=1}^{L}||z_l(\hat{\boldsymbol{x}})-\boldsymbol{u}_l(\hat{\boldsymbol{x}})||_{2,K}^2+\sum_{l=1}^{L}||q_l(\hat{\boldsymbol{x}})-p_l(\hat{\boldsymbol{x}})||_{1,K}^2\le Ce^{-bW} \tag{10}$$

holds true.

Remark 3.1 After obtaining a nonconforming solution, a set of corrections [10, 25] can be made such that velocity variable becomes conforming, and we have the error estimate

$$||\boldsymbol{u}-z||_{1,\Omega}+||p-q||_{0,\Omega}\le Ce^{-bW}. \tag{11}$$

Here, we briefly discuss the construction of the preconditioner. Using Theorem 2.1, we have

$$\mathcal{U}^{L,W}(\boldsymbol{u},p)\le C(\ln W)^2\mathcal{V}^{L,W}(\boldsymbol{u},p). \tag{12}$$

Using the trace theorem for Sobolev spaces, we get for some constant $\tilde{C}$

$$\tilde{C}\mathcal{V}^{L,W}(\boldsymbol{u},p)\le\mathcal{U}^{L,W}(\boldsymbol{u},p)\le C(\ln W)^2\mathcal{V}^{L,W}(\boldsymbol{u},p). \tag{13}$$

$\mathcal{U}^{L,W}(\boldsymbol{u}, p)$ is used as a preconditioner, so that the condition number of the preconditioned system is $O((ln W)^2)$, where W is the degree of the polynomial.

Residuals in the proposed scheme are computed by collocating the partial differential operator on a grid based on Gauss-Legendre-Lobatto points. Fractional Sobolev space norms are used to calculate jump along the interelement boundaries. Normal equations in the least-squares formulation are solved using the preconditioned conjugate gradient method.

4 Numerical Results

Here, we present the numerical results for Oseen equations on two-dimensional curvilinear domains and also Oseen and Navier-Stokes equations on three-dimensional domains. To show the performance of the proposed method, we have considered the Oseen and Navier-Stokes equations with Dirichlet boundary conditions whose exact solution is known and the approximate solution for different values of W is obtained. Let us denote the error between $\boldsymbol{u}$ and its approximate solution z in H^1 norm by $\|E_{\mathbf{u}}\|_1$, and the error between p and q in L^2 norm by $\|E_p\|_0$. $\|E_c\|_0$ denotes the error in continuity equation in L^2 norm, and it measures mass conserving property of the scheme.

4.1 *Ex-1: Oseen Equations on Annular Domain*

Consider the Oseen equation with $Re = 1$ and $\mathbf{b} = (1, 1)$ on the annular domain $\Omega = \{(r, \theta) : 1 \leq r \leq 4 \text{ and } 0 \leq \theta \leq \frac{\pi}{2}\}$ with Dirichlet boundary condition on the boundary.

The domain is divided into 4 curvilinear elements as shown in Fig. 1. Blending elements have been used [13]. The data is chosen such that

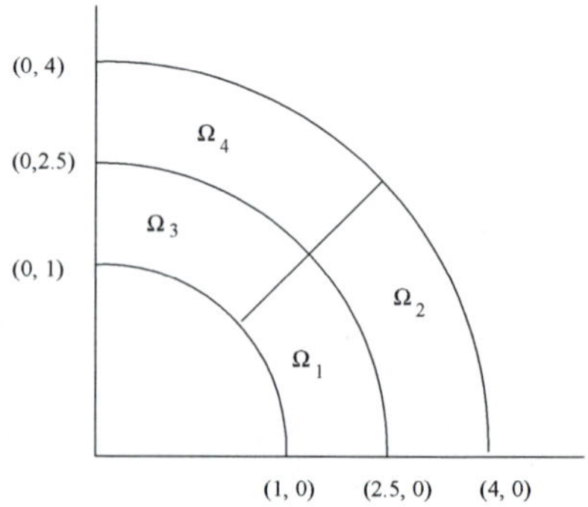

Fig. 1 Annular domain Ω and its discretization

Table 1 E_1, E_2 and $\|E_c\|_0$ for various values of W

W	E_1	E_2	$\|E_c\|_0$	itr
2	1.6863366E−01	9.4595302E−01	340.7201447E−00	12
3	6.2004035E−02	7.60763473E−02	148.0620495E−00	114
4	1.1500896E−02	1.66468232E−02	22.1068832E−00	177
5	2.6299851E−03	4.33600801E−03	7.32489306E−00	335
6	4.3727006E−04	6.29250872E−04	1.12044908E−00	457
7	4.9014479E−05	1.02811589E−04	1.59651009E−01	920
8	9.5979782E−06	6.69426435E−06	2.53922080E−02	1234
9	8.1496938E−07	1.39316542E−06	2.11950535E−03	1943
10	1.7114827E−07	1.14515414E−07	3.62696297E−04	2726

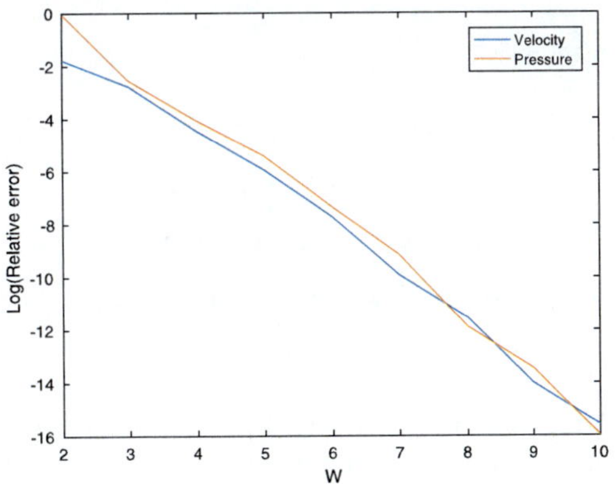

Fig. 2 Log of relative errors against W

$$u_1 = 20\, x_1\, x_2^3,$$
$$u_2 = 5(x_1^4 - x_2^4),$$
$$p = 60x_1^2 x_2 - 20x_2^3 + c$$

from an exact solution.

Table 1 shows the relative errors $E_1 = \frac{\|E_{\mathbf{u}}\|_1}{\|\mathbf{u}\|_1}$, $E_2 = \frac{\|E_p\|_0}{\|p\|_0}$ and $\|E_c\|_0$ and the number of iterations for various values of W. Figure 2 shows the log of the relative errors against W. The curve is almost linear, and this shows that the error decays exponentially in $\|E_{\mathbf{u}}\|_1$ and $\|E_p\|_0$ norms.

4.2 Ex-2: Oseen Equations on Three-Dimensional Domains

Consider the Oseen equations on $[-1, 1]^3$ with $Re = 1$ and $\boldsymbol{b} = (1, 1, 1)$. The force function and boundary data are chosen such that the exact solution to the problem is

Table 2 Numerical results for Oseen equations on $[-1, 1]^3$

W	$\|E_u\|_1$	$\|E_p\|_0$	$\|E_c\|_0$	itr
2	1.2282E+01	1.7038E+01	1.9625E+00	13
3	1.6627E+00	2.7362E+00	3.7029E−01	51
4	9.5849E−03	1.9419E−02	5.9284E−03	122
5	1.8202E−04	5.1715E−04	9.7752E−05	341
6	2.0729E−05	2.0786E−04	1.1582E−05	575
7	3.1522E−06	1.5777E−05	1.6626E−06	858
8	3.9872E−07	7.9937E−07	2.2938E−07	1149
9	5.3780E−08	4.2063E−08	2.6611E−08	1352
10	3.2513E−09	5.6004E−09	2.1754E−09	2193

given by

$$\begin{aligned}
u_1(x_1, x_2, x_3) &= 4x_1^2 x_2 x_3 (1 - x_1)^2 (1 - x_2)(1 - x_3)(x_3 - x_2) \\
u_2(x_1, x_2, x_3) &= 4x_1 x_2^2 x_2 x_3 (1 - x_1)(1 - x_2)^2 (1 - x_3)(x_1 - x_3) \\
u_3(x_1, x_2, x_3) &= 4x_1 x_2 x_3^2 (1 - x_1)(1 - x_2)(1 - x_3)^2 (x_2 - x_1) \\
p(x_1, x_2, x_3) &= -2x_1 x_2 x_3 + x_1^2 + x_2^2 + x_3^2 + x_1 x_2 + x_1 x_3 + x_2 x_3 \\
&\quad - x_1 - x_2 - x_3.
\end{aligned}$$

Here, we consider a single element, i.e. $[-1, 1]^3$. Table 2 shows the errors $\|E_{\mathbf{u}}\|_1$, $\|E_p\|_0$ and $\|E_c\|_0$ and the number of iterations for various values of W.

4.3 Ex-3: Navier-Stokes Equations on Three-Dimensional Domains

Navier-Stokes equations on a domain $\Omega \subset \mathbb{R}^3$ are given by

$$\begin{aligned}
-\Delta \boldsymbol{u} + (\boldsymbol{u}.\nabla)\boldsymbol{u} + \nabla p &= \boldsymbol{f} \\
-\nabla.\boldsymbol{u} &= h \\
\boldsymbol{u}|_{\partial\Omega} &= \boldsymbol{g}.
\end{aligned} \tag{14}$$

To obtain the solution of the Navier-Stokes equations, we solve a sequence of Oseen type equations

$$-\Delta \boldsymbol{u}^n + (\boldsymbol{u}^{n-1}.\nabla)\boldsymbol{u}^n + \nabla p^n = \boldsymbol{f} \tag{15}$$

Table 3 Numerical results for Navier-Stokes equations on $[-1, 1]^3$

W	$\|E_{\mathbf{u}}\|_1$	$\|E_p\|_0$	$\|E_c\|_0$
2	1.0788E+01	2.4904E+00	2.6904E+00
3	5.9363E−01	9.0542E−01	3.2288E−01
4	8.6318E−02	1.0882E−01	6.5594E−02
5	2.3597E−03	1.4203E−02	1.8692E−03
6	7.1142E−04	7.7410E−04	5.3313E−04
7	5.2635E−05	6.1185E−05	4.1795E−05
8	1.9171E−08	3.8987E−08	1.2057E−08

and use the fixed point iteration technique. The initial guess $(\boldsymbol{u}^0, p^0)$ will be obtained by Stokes solver, and the sequence of approximations $(\boldsymbol{u}^n, p^n), n = 1, 2, ..$ are obtained by solving the Oseen type of equations given Eq. (15).

Here, we consider Navier-Stokes equations on the domain $[-1, 1]^3$ with Dirichlet boundary condition on the boundary. The force function and boundary data are chosen such that the exact solution of the problem is the same as the solution in example 2. We consider only one element and obtained the approximate solution. Stopping criteria based on the difference between solutions from two consecutive iterations is used to achieve convergence. Table 3 shows the errors $\|E_{\mathbf{u}}\|_1$, $\|E_p\|_0$ and $\|E_c\|_0$ for various values of W.

5 Conclusion

Here, we have studied the nonconforming spectral element method for Oseen equations and Navier-Stokes equations. Same order spectral element functions have been used for both velocity and pressure variables. Numerical results have been presented for Oseen equations on a curvilinear domain in two dimensions and for Oseen and Navier-Stokes equations in three-dimensional domains. Numerical results confirm exponential accuracy.

References

1. Agmon S, Douglis A, Nirenberg L (1964) Estimates near the boundary for solutions of elliptic partial differential equations satisfying general boundary conditions II. Commun Pure Appl Math 17:35–92
2. Aziz AK, Kellog RB, Stephens AB (1985) Least squares methods for elliptic systems. Math Comput 44(169):53–70
3. Bochev PB (1997) Analysis of least-squares finite element methods for the Navier-Stokes equation. SIAM J Numer Anal 34(5):1817–1844
4. Bochev PV, Gunzburger MD (2009) Least-squares finite element methods. Springer

5. Burger M (2010) Numerical methods for incompressible flows. Lecture notes. https://ww3.math.ucla.edu/camreport/cam04-12.pdf
6. Cesmelioglu A, Cockburn B, Nguyen NC, Peraire J (2013) Analysis of HDG methods for Oseen equations. J Sci Comput 55:392–431
7. Cockburn B, Kanschat G, Schotzau D (2003) The local discontinuous Galerkin method for the Oseen equations. Math Comput 73:569–593
8. Deang JM, Gunzburger MD (1998) Issues related to least-squares finite element methods for the Stokes equations. SIAM J Sci Comput 20:878–906
9. Duan HY, Liang GP (2003) On the velocity-pressure-vorticity least-squares mixed finite element method for the 3D Stokes equations. SIAM J Numer Anal 41(6):2114–2130
10. Dutt PK, Kumar NK, Upadhyay CS (2007) Non-conforming $h-p$ spectral element methods for elliptic problems. Proc Indian Acad Sci (Math Sci) 117(1):109–145
11. Eason ED (1976) A review of least-squares methods for solving partial differential equations. Int J Math 10:1021–1046
12. Franca LP, John V, Matthies G, Tobiska L (2007) An inf-sup stable and residual-free bubble element for the Oseen equations. SIAM J Numer Anal 45:2392–2407
13. Gordan WJ, Hall CA (1973) Transfinite element methods: blending-function interpolation over arbitrary curved element domains. Numer Math 21(2):109–129
14. Heinrichs W (2004) Least-squares spectral collocation for the Navier-Stokes equations. J Sci Comput 21:81–90
15. Hessari P, Shin BC (2013) The least-squares pseudo-spectral method for Navier-Stokes equations. Comput Math Appl 66(3):318–329
16. Husain A (2011) $h-p$ spectral element methods for three dimensional elliptic problems on non-smooth domains using parallel computers. PhD thesis, IIT Kanpur India. Reprint available at http://arxiv.org/abs/1110.2316
17. Jiang BN, Sonnad V (1994) Least-squares solution of incompressible Navier-Stokes equations with p-version of finite elements. Comput Mech 15:129–136
18. Jiang BN (1997) The least-squares finite element method. Springer
19. John V (2016) Finite element method for incompressible flow problems. Springer
20. Kim SD, Lee CO, Manteuffel TA, Mccormick SF, Rohrle O (2006) First order system least-squares for the Oseen equations. Numer Linear Algebra Appl 13(7):523–542
21. Mohapatra S, Ganesan S (2016) Non-conforming least squares spectral element formulation for Oseen equations with applications to Navier-Stokes equations. Numer Funct Anal Optim 37(10):1295–1311
22. Pontaza JP, Reddy JN (2003) Spectral/hp least-squares finite element formulation for the Navier-Stokes equations. J Comput Phys 190:523–549
23. Proot MMJ, Gerritsma MI (2002) A least-squares spectral element formulation for the Stokes problem. J Sci Comput 17:285–296
24. Proot M, Gerritsma MI (2002) Least-squares spectral elements applied to the Stokes problem. J Comput Phys 181:454–477
25. Schwab Ch (1988) p and $h-p$ finite element methods. Clarendon Press, Oxford
26. Temam R (1977) Navier-Stokes equations, theory and numerical analysis. North-Holland Publishing Company, New York
27. Tsai CC, Yang SY (2005) On the velocity-vorticity-pressure least-squares finite element method for the stationary incompressible Oseen problem. J Comput Appl Math 182:211–232

Application of Gorilla Troops' Social Intelligence in Damage Detection for a Girder Bridge

Long Viet Ho, Thanh Bui-Tien, and Magd Abdel Wahab

Abstract Structural damage diagnosis employing optimization techniques has been receiving the attention of scientists worldwide. This is due to the simplicity of implementing a stochastic optimization process and the robust development of optimization algorithms. These advantages create diverse applications of optimization algorithms in many areas of life. An inverse problem-based approach for damage detection is very time-consuming. Artificial neural networks (ANN) can overcome this drawback. However, ANN's performance is much dependent on its architecture. Recently, a heuristic optimization algorithm, Artificial Gorilla troops optimizer (AGTO), was developed. Its effectiveness has been proved through optimization benchmarks and engineering problems. In this paper, Gorilla troop's social intelligence is utilized to identify damaged segments in a simply supported girder bridge. AGTO serves as a trainer in supervised learning to determine the optimized number and size of hidden layers in the architecture of ANN. First, the bridge's finite element model is modelled using as-built drawings. Then, each girder of the bridge is assumed to be suffered stiffness reduction at several locations to simulate single and multi-damage. The frequency changes and mode shapes are the inputs of ANN. The error in damage location and severity between prediction and target is the objective function of the proposed approach. Several types of objective functions, like mean squared error (MSE), root-mean-square error (RMSE), mean absolute error (MAE) and standard deviation-based function (SDBF), are investigated for this problem. The findings confirm that the computational cost of the proposed approach is more significant

L. V. Ho (✉)
Division of Bridge Engineering and Underground Infrastructure, Faculty of Civil Engineering, University of Transport and Communications, Campus in Ho Chi Minh City, Ho Chi Minh, Vietnam
e-mail: longhv_ph@utc.edu.vn

T. Bui-Tien
Department of Bridge Engineering and Underground Infrastructure, Faculty of Civil Engineering, University of Transport and Communications, Hanoi, Vietnam

M. Abdel Wahab
Department of Electromechanical, Systems and Metal Engineering, Faculty of Engineering and Architecture, Ghent University, Ghent, Belgium

M. Abdel Wahab (ed.), *Proceedings of the 5th International Conference on Numerical Modelling in Engineering*, Lecture Notes in Civil Engineering 311,
https://doi.org/10.1007/978-981-19-8429-7_2

than ANN's. In contrast, the accuracy and precision of the AGTOANN are superior. Besides, in light of prediction accuracy, MAE and RMSE show better performance compared with MSE and SDBF.

Keywords Dynamic properties · AGTOANN · Damage detection

1 Introduction

Vibration data is crucial because they help to identify warning signs of potential failures in a structure or a bridge. Damage sources are diverse, for instance, overloading, environmental conditions, chemical attack, material ageing, etc. Early structural damage detection ensures a long service life of the structure. Any presence of damage that affects the structural stiffness will result in a change in the dynamic characteristics. Through analysing shifts in these modal properties, the bridge manager can point out possible failures in the structure being monitored. The inverse problem-based approach and artificial neural network (ANN) are two common methods used in fault diagnosis. However, the former approach is based on an iterative process, so making a diagnosis requires much time [1–3]. In contrast, the latter is based on a training process of the network. With new data, the trained network can immediately make decisions with high accuracy. Therefore, this approach proves to be very effective for real-time structural monitoring.

However, using ANN in fault diagnosis needs to address some issues related to the hidden layer. The number of hidden layers is a common research question. The choice of this number can be affected by the complexity and dimensions of a training dataset. Some researchers proved that one hidden layer could be successfully applied to their problems [4–6]. Other studies investigated enhancing the performance of ANN using more than one hidden layer. These authors used two hidden layers [7, 8] or a maximum of three hidden layers [9]. However, the number of hidden layers in these studies was based on experience or investigated manually. Furthermore, the time complexity is proportional to the number of hidden layers in ANN because the network needs more training time.

The size of a hidden layer or the number of hidden neurons is also a controversial issue with many different points of view. The choice of size of the hidden layer could significantly impact the convergence rate and the accuracy of the model. Some studies selected the number of hidden neurons based on experience or manual investigation [6, 10]. Authors in [11–13] recommended that the number of hidden nodes should be selected concerning the number of input and output nodes. The same [7, 9] and different [8] numbers of hidden nodes for each layer were also investigated.

Apart from these aforementioned issues, the evaluation of a trained network received attention from the research community. Evaluation metrics are used to measure or quantify how well a predictive model fits a dataset. For evaluating the performance of a regression model, a common metric, namely mean squared error (MSE), is often used [14, 15]. Other authors use the root-mean-square error (RMSE) in their research [16]. Mean absolute error (MAE) demonstrated its advantage over RMSE in [17]. In contrast, authors in [18] confirmed that RMSE could gain more benefits than MAE. The author in [19] conducted a comparison study to find out the properties of MAE and RMSE based on different error distributions. The vectorization technique also was used as an evaluation metric. This technique was applied successfully for damage detection in [20]. Besides, a reduction in the dimension of a dataset can decrease the computational time of the training process. It can be seen that a different evaluation metric fits a different set of regression models. No unique value matches every problem or dataset. For this reason, determining the mentioned issues is essential to ensure the effectiveness of ANN's performance, especially in technical problems.

Therefore, this current study employs a recent algorithm, Artificial Gorilla troops optimizer (AGTO) [21], to optimize the size and number of hidden layers associated with several objective functions. MSE, RMSE, MAE and standard deviation-based function[1] (SDBF) are investigated. The optimized values of training parameters for ANN are used to identify damage in a girder bridge. In addition, a comparison study is conducted to point out a suitable evaluation metric for ANN. The paper is organized into four sections. The first section is the Introduction. Section 2 presents the methodologies of the proposed approach. The applications of the method in failure identification are in Sect. 3. The last section shows the main findings of this study.

2 Methodologies

The core idea of this study is to employ the global optimization capabilities of evolutionary optimization algorithms to determine the best combination of training parameters for ANN. However, different evaluation metrics are investigated in this study to evaluate the effectiveness of training the network.

2.1 Enhancing ANN Using AGTO (AGTOANN)

Artificial gorilla troop optimizer (AGTO) [21] was inspired by gorillas' life in nature. A silverback gorilla is in charge of leading its troop to find food together, making a decision about group movements, and the group's safety. It can be said that the silverback is the heart of a troop. Therefore, AGTO mimicked the gorilla behaviours via a mathematical mechanism for optimization operations. In the AGTO algorithm, the silverback gorilla represents the best solution so far.

In this study, AGTO is employed to identify the optimized values of size and number of hidden layers. Therefore, these parameters are treated as investigated

[1] This function is calculated based on the vectorization technique, please see Ref. [20].

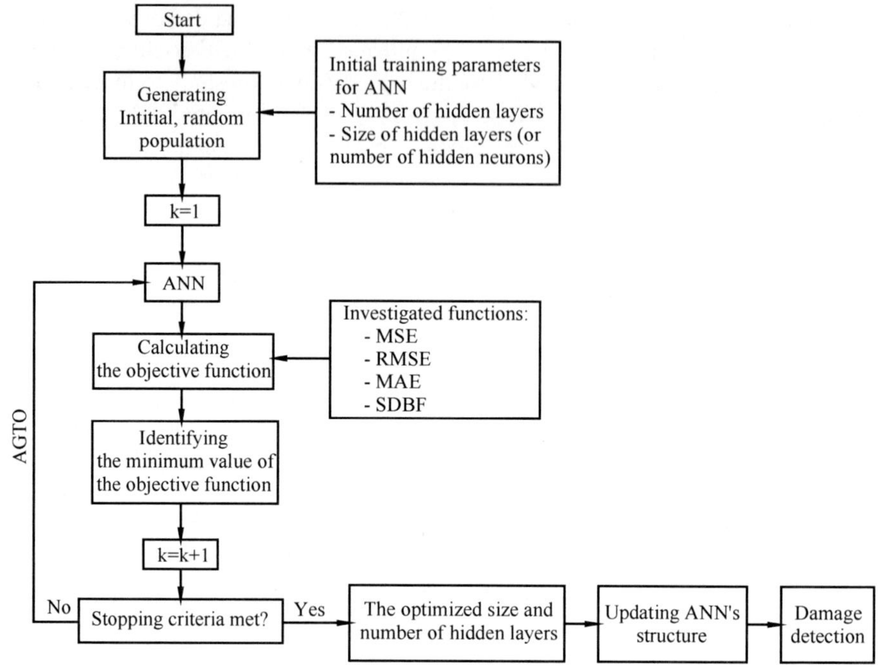

Fig. 1 Step-by-step procedure of using AGTO to optimize the architecture of ANN

variables in AGTO. In the first step, an initial population of gorillas is generated randomly to kick off an optimization process. Each gorilla consists of size and number of hidden layers. Then the fitness value of each particle is calculated by ANN. After each iteration, the best fitness is assigned to a silverback gorilla. The silverback gorilla holds a set of the size and number of hidden layers that creates the minimum deviation between the predicted and actual values. The process repeats until meeting a stop condition, such as maximum iteration. The silverback gorilla, which contains the optimized values of the hidden neurons, and hidden layers, is returned at the final iteration. A work flowchart of the proposed approach, AGTOANN, is presented in Fig. 1.

2.2 *Evaluation Metrics*

a. Mean squared error (MSE)

MSE is a popular metric which is used in ANN. Due to the square of the error, MSE tends to exaggerate the mean difference between the actual and predicted values. The lower MSE value implies that a model and dataset match well.

$$MSE = \frac{1}{n}\sum_{i=1}^{n}\left(y_{predicted} - y_{actual}\right)^2 \quad (1)$$

b. Root-mean-square error (RMSE)

The calculation is similar to the MSE, but the RMSE reduces the exaggeration of errors by using square roots. In other words, RMSE is the square root of MSE.

$$RMSE = \sqrt{\frac{\sum_{i=1}^{n}\left(y_{predicted} - y_{actual}\right)^2}{n}} \quad (2)$$

c. Mean absolute error (MAE)

MAE is the most straightforward metric that computes the absolute error between the estimated and actual values. Then all these errors are summed up and divided by an overall number of samples.

$$MAE = \frac{1}{n}\sum_{i=1}^{n}\left|y_{predicted} - y_{actual}\right| \quad (3)$$

d. Standard deviation-based function (SDBF)

This function is calculated based on vectorization technique [20]. Details of the calculation of this objective function are in Eqs. (4–7).

$$SDBF = \sqrt{\sum_{i=1}^{n}\left[(V_i)^2 - \left(\overline{V}_i\right)^2\right]} \quad (4)$$

$$V_i = \frac{1}{n-1}\sum_{i=1}^{n}\left|y_{actual}^{i} - \overline{y}_{actual}\right|^2 \quad (5)$$

$$\overline{V}_i = \frac{1}{n-1}\sum_{i=1}^{n}\left|y_{predicted}^{i} - \overline{y}_{predicted}\right|^2 \quad (6)$$

$$\overline{y} = \frac{1}{n}\sum_{i=1}^{n} y_i \quad (7)$$

where n is the total number of samples, $y_{predicted}$, y_{actual}, $\overline{y}$ indicate estimated, actual and average values, respectively, and V_i, $\overline{V}_i$ imply vectorization matrices of actual and predicted values.

3 Application of Proposed Approach in Damage Detection

3.1 Bridge Description

Doan Hung bridge ensures interprovincial traffic between Ha Giang–Tuyen Quang with Phu Tho and Vinh Phuc provinces in National Highway No. 2. An overview of the bridge is introduced in Fig. 2. Doan Hung is a five-span simply supported bridge with four reinforcement piers and two abutments. Each span consists of four prestressed concrete T-shaped girders (see Fig. 4). The length of each span varies from 23.93 to 24.25 m (see Fig. 3). The FE model of the bridge was built based on as-built drawings before it was retrofitted by external prestressing. Because of the similar boundary conditions with steel-laminated elastomeric bearings, the centre span is chosen to investigate in this study.

Fig. 2 Overview of Doan Hung bridge, a five-span simply supported girder bridge

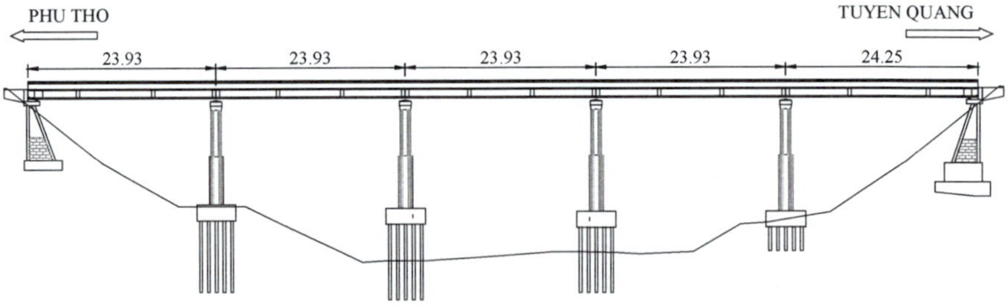

Fig. 3 Configuration of the bridge (Unit: m)

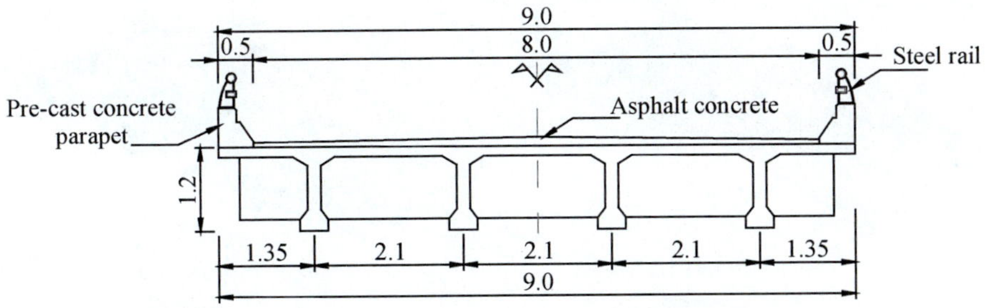

Fig. 4 Bridge cross-section at midspan (Unit: m)

3.2 Finite Element Model

Solid185 element in ANSYS [22] was used to build the FE model of the simply supported girder bridge. Mesh properties of the FE model of the bridge consisted of tetrahedral elements, free mesh due to complex geometries and element sizes 0.1 and 0.4 m for diaphragm beams and T-shaped girders, respectively. Material properties in the FE model are shown in Table 1. Other non-structural components, such as steel bridge rail and asphalt wearing surface, were considered as added masses in the FE model (see Fig. 5). Flange-to-flange connection between two Tee girders was fully rigid.

Table 1 Mechanical properties of concrete material for the FE model

No	Components	Young's modulus E (GPa)	Density ρ (kg/m^3)	Poisson's ratio ν
1	T girders, precast bridge parapets	36.05	2600	0.2
2	Cross-beams	31.22	2600	0.2

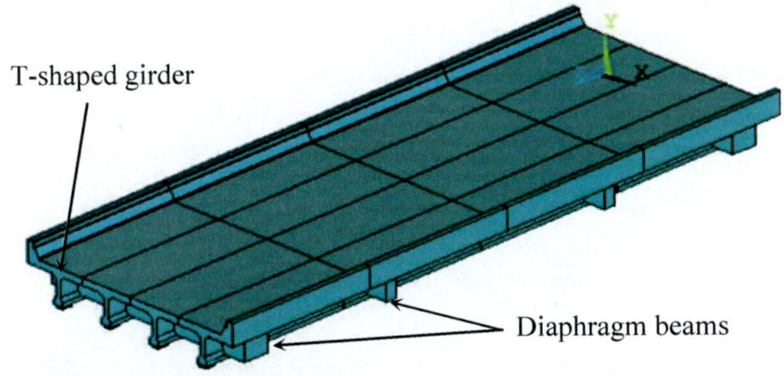

Fig. 5 FE model of the centre span

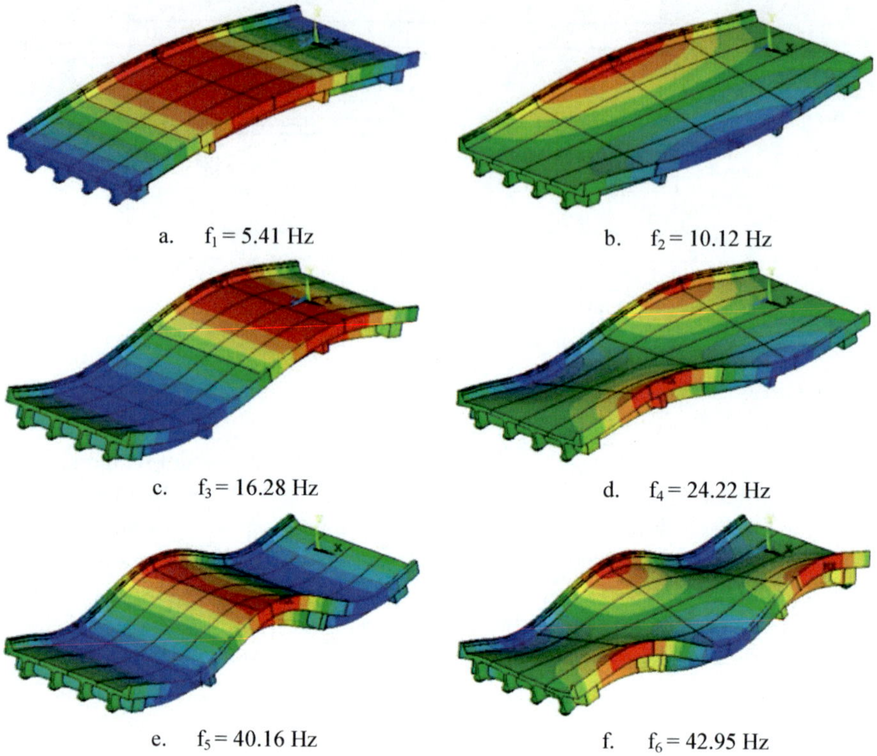

Fig. 6 The first six modes of the bridge (**a** 1st vertical bending mode, **b** 1st torsion mode, **c** 2nd vertical bending mode, **d** 2nd torsion mode, **e** 3rd vertical bending mode, **f** 3rd torsion mode)

3.3 *Modal Analysis*

Block Lanczos method was employed to perform the free modal analysis of the bridge. Results of the modal analysis are shown in Fig. 6. Dynamic properties of the first six modes, such as frequencies and mode shape data at a 36-point grid (see Fig. 7 and Fig. 8), were collected. Then, these data were utilized to calculate an objective function of the hybrid model, AGTOANN, for damage identification.

3.4 *Damage Scenarios*

The current study focused on detecting local damage in the main girder. Therefore, four segments in each girder were used to simulate the failures. Figure 9 introduces the 16 segments that were used to represent failures in the girder bridge. In this

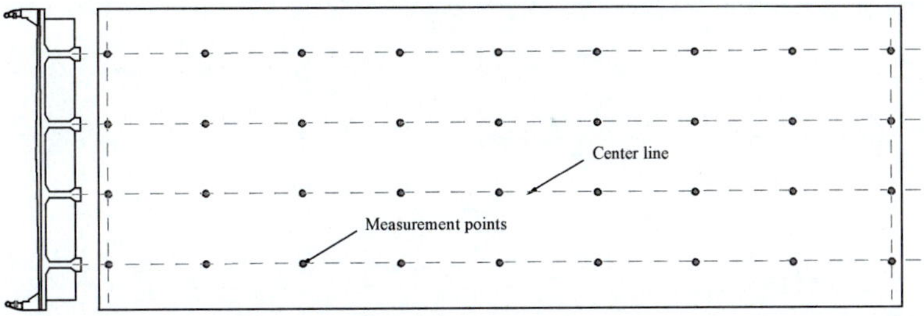

Fig. 7 Top-plane view: A 9 × 4 measurement grid at four centre lines of the girders

Fig. 8 The mode shape data of the second mode using the 36-point grid

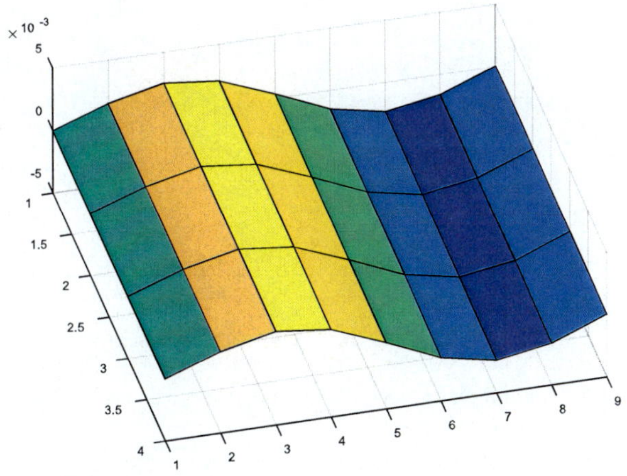

section, two damage scenarios, including single damage and multiple damages, were generated to verify the efficiency of the proposed approach.

The local failures were simulated by an assumption that the bending stiffness of each segment was reduced. Therefore, in the first damage scenario, each segment suffers a bending stiffness reduction in an interval [1% ÷ 50%] by a step of 1%. This hypothesis created $C_{16}^{1} \times 50 = 800$ training samples. In the second scenario, the stiffness reduction from 1 to 40% was applied to random couples of segments out of 16 ones. Samples of training data were $C_{16}^{2} \times 40 = 4800$. Besides, nineteen cases for each damage scenario were generated to verify the effectiveness of the trained networks.

3.5 ANN's Structure

As discussed in the introduction, the number of hidden layers and the number of hidden neurons in each layer impact the performance of ANN. Therefore, these parameters should be optimized in light of the computational time and the accuracy.

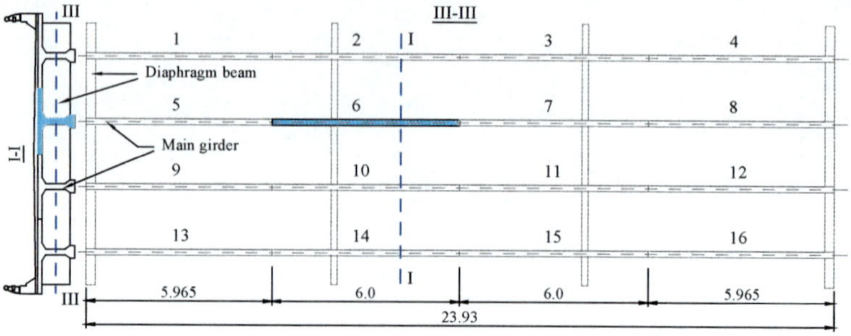

a. Single damaged segement, 6

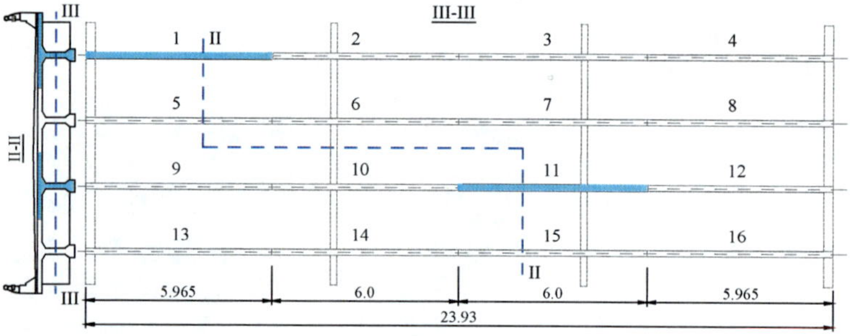

b. Two damaged segments, 1 and 11.

Fig. 9 Bottom-plane view: Illustration of two damage scenarios and segment labels

Table 2 summarises these training parameters for the traditional ANN and the hybrid model, AGTOANN. Besides, some initial parameters for an optimization process, like the number of populations and the maximum iteration, are 30 and 15.

Table 2 Initial ANN's architecture

Methods	S_{input}	N_{Hidden}	S_{Hidden}	S_{output}	Objective function
ANN	12	1	$2/3 \times N_{input} + N_{output}$	2 or 4	MSE
AGTOANN (P = 30, It = 15)	12	$1 \div 2$	$2 \div 2 \times N_{input}$	2 or 4	MSE, RMSE, MAE, SDBF

where S_{input}, S_{Hidden}, S_{output} indicate the number of neurons in input, hidden and output layer, respectively. N_{Hidden} implies the number of hidden layers. P and It indicate the maximum number of population and the number of iterations, respectively.

Table 3 The architecture of ANN for the single-damage scenario

Methods	Number of hidden layer	Size of hidden layer 1	Size of hidden layer 2
ANN	1	10	
AGTOANN-MSE	2	13	12
AGTOANN-RMSE	2	9	17
AGTOANN-MAE	2	20	17
AGTOANN-SDBF	1	17	

3.6 *Results of the Training Process*

a. Single-damage identification

During the stochastic optimization process, AGTO tried to look for the best silverback gorilla among other potential gorillas based on the fitness values. In the final iteration, the choice of the best silverback gorilla implied the optimized values of the number and size of hidden layers. Table 3 introduces the structure of ANN associated with several approaches. For example, AGTOANN-SDBF proposed using one hidden layer, while the others identified two layers for the first damage scenario.

These updated parameters were then input to ANN for damage identification. Results of the training process are displayed in Table 4, Figs. 10, 11. Compared with ANN, the proposed approach shows better performance due to smaller values of MSE. It can be seen that the evaluation metric, MAE, obtained the best MSE value in the training process. While the two metrics, RMSE and MSE, perform stability in all stages from training, validation and test. The RMSE generally achieved the best MSE value compared to the rest metrics.

In this case, two performance parameters, regression analysis and error histogram were chosen to assess the quality of the trained network. Firstly, a good agreement between the predicted and actual values is performed in Fig. 10. Data were allocated vicinity of the 45-degree line when ANN was used. The proposed approach revealed a much better training process because almost data perfectly fit the diagonal line. The correlation coefficient R-value, which the hybrid model achieved, was higher than that of ANN. Secondly, Fig. 11 shows the smallest errors between predicted and

Table 4 Comparative parameters in the single-damage scenario

Parameters	ANN	AGTOANN			
	MSE	MSE	RMSE	MAE	SDBF
CPU-time (s)	5.5	3,126	4,112	4,034	1,344
MSE-Training	3.72E-01	5.84E-03	4.08E-04	3.92E-05	5.88E-02
MSE-Validation	2.83E-01	1.19E-02	2.37E-02	7.10E-02	6.94E-01
MSE-Test	4.44E-01	2.47E-02	7.76E-03	1.52E-02	3.05E-01
MSE-All	0.3693	0.0096	0.005	0.013	0.191

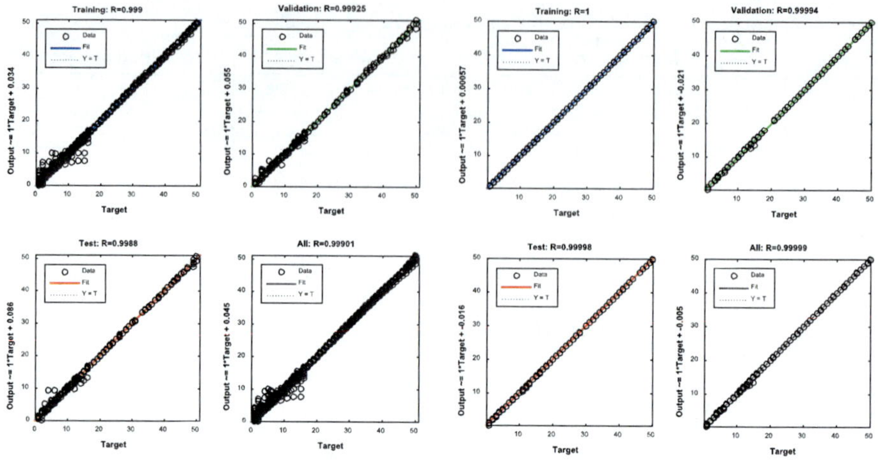

ANN AGTO-ANN using RMSE

Fig. 10 Regression plots for damage scenario 1

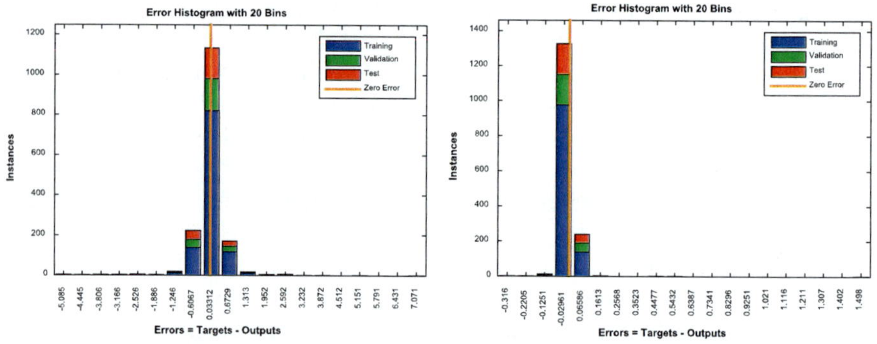

ANN AGTO-ANN using RMSE

Fig. 11 Error histogram for damage scenario 1

desired values, 0.02961 and 0.03312, for the hybrid model and ANN, respectively. The number of instances in the error histogram using AGTOANN was over 1300, larger than that of ANN, approximately 1100.

As mentioned in Sect. 3.4, nineteen single scenarios were randomly generated to evaluate the trained network. It should note that these new data are not in the training dataset. The damage identification results are revealed in Table 5. For an intuitive view, Fig. 13 presents results from Table 5. In this figure, the values below the zero point imply that the model fails to identify the damage location. The findings indicate that the proposed approach has superior performance over ANN. It could identify exactly almost damaged segments. Only one misidentified location was produced

Table 5 Results of failure identification in the single-damage scenario

Actual Damage		ANN		AGTOANN							
		MSE		MSE		RMSE		MAE		SDBF	
Elem	Level	Elem	Level	Elem	Level	Elem	Level	Elem	Level	Elem	Level
6	43.3	7	43.19	6	43.32	6	43.32	6	43.30	6	43.25
10	39.2	10	39.29	10	39.22	10	39.21	10	39.20	10	39.26
16	34.3	16	34.26	16	34.32	16	34.28	16	34.30	16	34.38
3	39	3	39.16	3	39.05	3	38.99	3	39.00	3	39.01
14	8.4	13	8.30	14	8.33	14	8.40	14	8.40	14	8.41
1	36.3	1	36.05	1	36.26	1	36.27	1	36.31	1	36.23
13	9.8	13	9.61	13	9.83	13	9.81	13	9.79	13	9.84
4	30.3	4	29.95	4	30.28	4	30.31	4	30.31	4	30.27
5	23.7	5	23.81	5	23.69	5	23.70	5	23.70	5	23.66
2	43.3	2	43.45	2	43.30	2	43.30	2	43.30	2	43.31
7	11.7	6	11.81	7	11.72	7	11.69	7	11.70	7	11.76
15	4	14	3.76	15	3.93	15	4.03	15	3.99	15	4.01
9	36.6	9	36.79	9	36.60	9	36.60	9	36.60	9	36.66
9	31.7	9	31.92	9	31.72	9	31.71	9	31.70	9	31.69
14	39.4	13	39.31	14	39.37	14	39.39	14	39.40	14	39.38
13	32.1	14	32.75	13	32.07	13	32.08	13	32.10	13	32.11
3	41.3	2	41.47	3	41.34	3	41.30	3	41.30	3	41.31
12	38.8	11	38.67	12	38.76	12	38.80	12	38.79	12	38.75
2	7.6	1	7.83	2	7.62	2	7.60	2	7.60	3	7.68

when SDBF was used. Besides, the proposed approach also predicted severities closer to the actual ones than using the traditional ANN (see Fig. 13).

The two-evaluation metrics, RMSE and MAE, showed their dominance in damage quantification over the others, e.g. MSE and SDBF. The predicted values are pretty similar to the actual ones. The absolute errors are less than 0.05. Details of several failure identifications are plotted in Fig. 12.

b. Multi-damage identification

In the second scenario, the two metrics, RMSE and MAE, were used to detect failures in the observed structure. The stochastic optimization achieved the same outcome of ANN's structure as in Table 7 for the two mentioned objective functions, RMSE and MAE. The hybrid model continues producing a smaller MSE value, 0.29, than that of ANN, 2.5 (see Table 6). Then, a 1–2–4 architecture of ANN was utilized to localize and quantify the damage.

The findings confirm that AGTO can work well with proposing an optimized architecture for ANN based on training data. The confirmation is based on The higher R-value of the correlation coefficient (Fig. 14), the higher number of zero

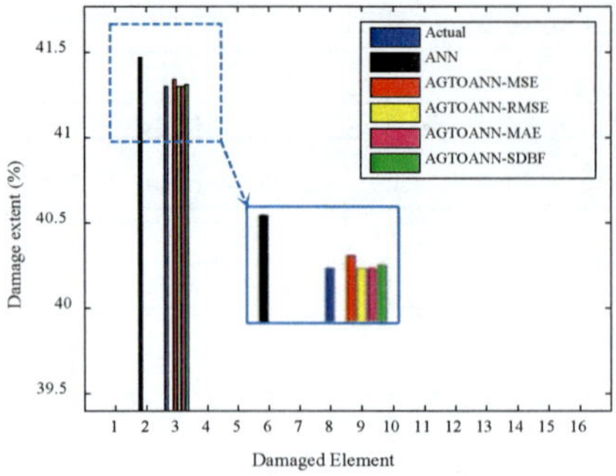

a. 41.3% at element 3

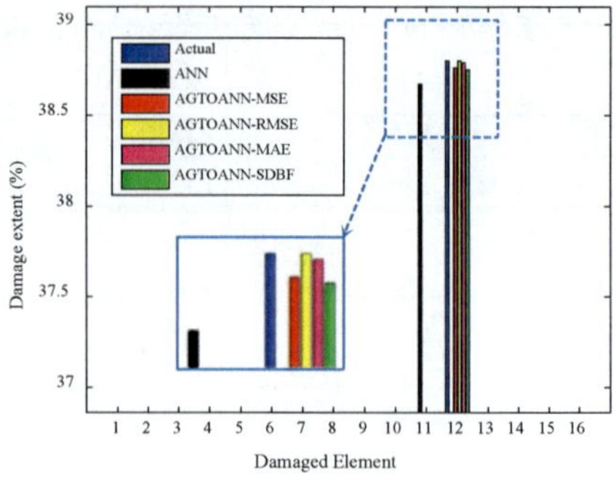

b. 38.8% at element 12

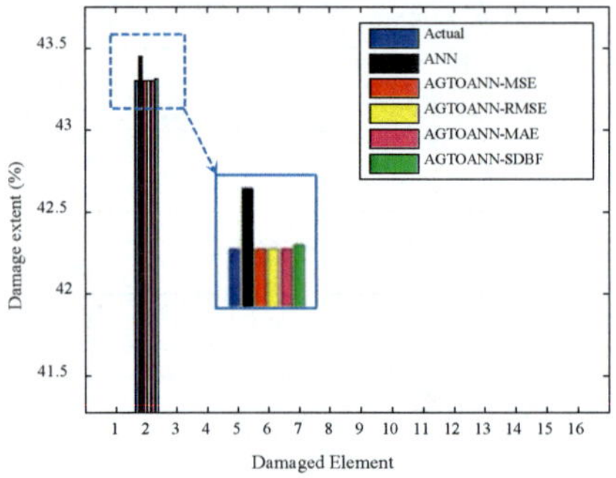

c. 43.3% at element 2

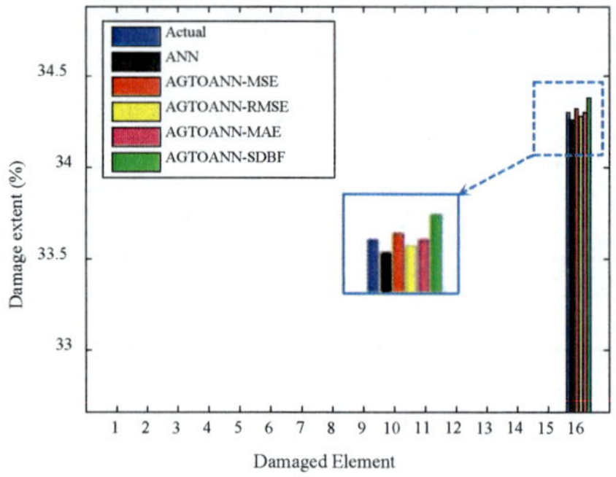

d. 34.3% at element 16

Fig. 12 Single-damage identification using ANN and AGTOANN

Fig. 13 Summary of damage identification results of the 19 test cases

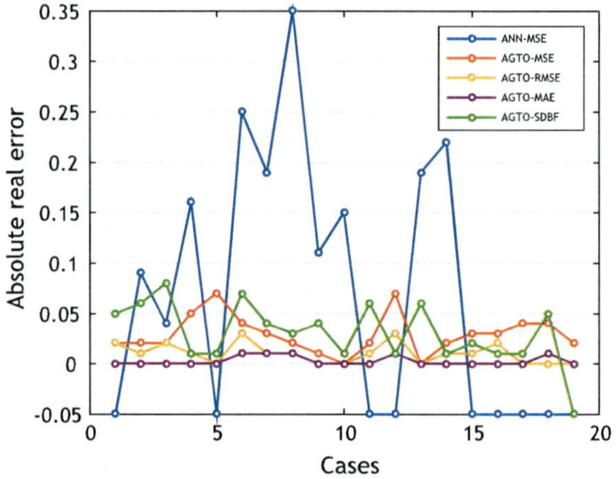

Table 6 Comparative parameters in the two-damage scenarios

Parameters	ANN	AGTOANN	
	MSE	RMSE	MAE
CPU-time (s)	11.4	12,681	12,615
MSE-Training	2.487	0.242	0.242
MSE-Validation	2.618	0.389	0.389
MSE-Test	2.448	0.413	0.413
MSE-All	2.501	0.2895	0.2895

Table 7 The optimized architecture of ANN for multi-damage scenario

Methods	Number of hidden layer	Size of hidden layer 1	Size of hidden layer 2
ANN	1	12	
AGTOANN	2	23	24

error values (Fig. 15) and more minor errors between targets and predicted values (Fig. 16).

A dataset of 19 cases evaluated the efficiency of the trained network. Table 8 performs all the results of damage identification. For comparison purposes, these results are plotted in Figs. 17, 18. In each case, if both failure locations are correctly determined, it is counted as the correct prediction. However, if only 1 location is correctly identified, it is still considered the wrong forecast. For example, the outcomes in Fig. 17 indicate that ANN could successfully localize 2 out of 19 cases.

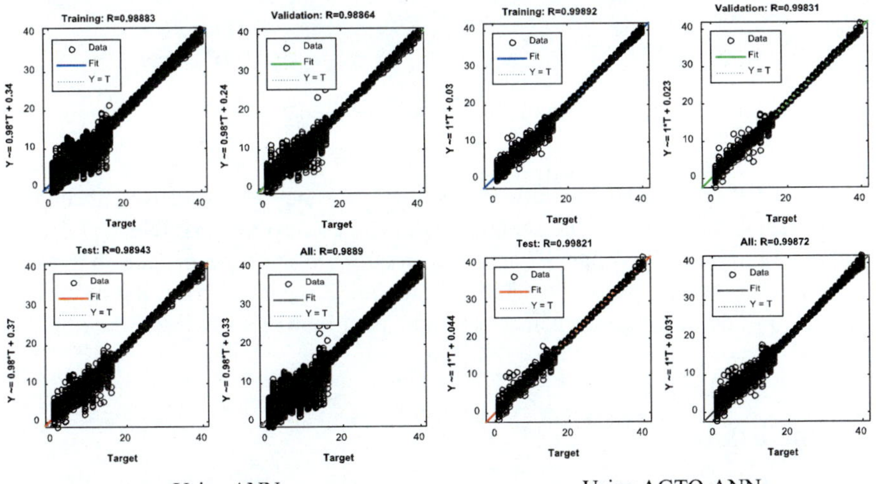

Fig. 14 Regression plots for damage scenario 2

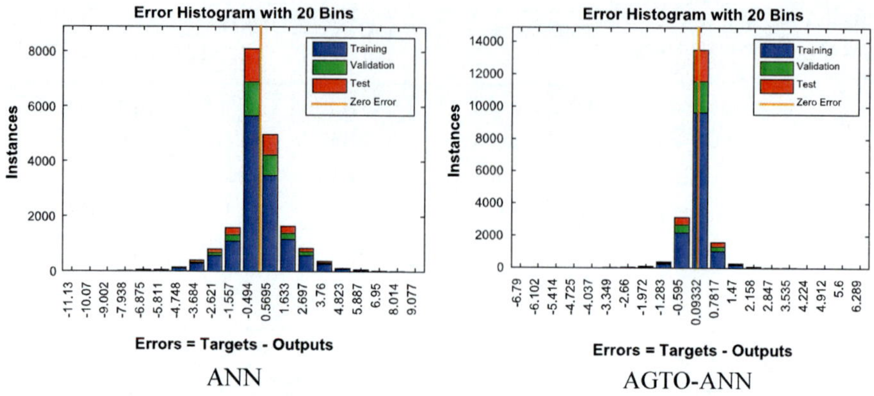

Fig. 15 Error histogram for damage scenario 2

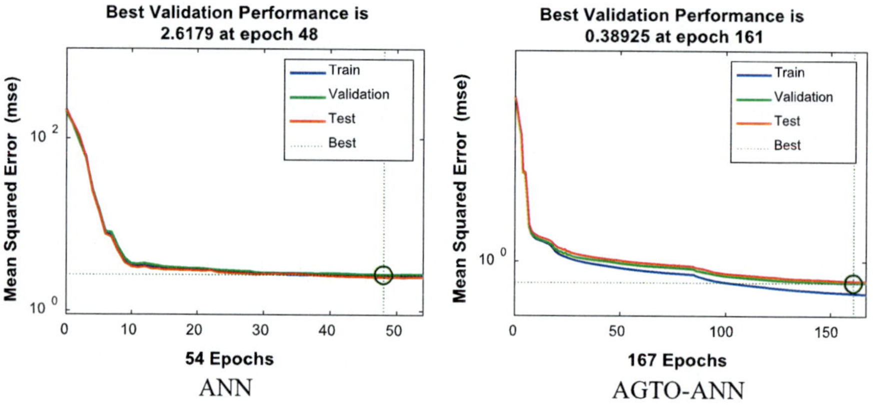

Fig. 16 Network training performance plots for damage scenario 2

Meanwhile, the proposed model failed to identify only one case. Moreover, the minor absolute real errors could not be achieved using the conventional ANN.

In general, using the proposed approach requires more time in fault diagnosis than using a traditional ANN (see Table 4 and Table 6). In contrast, the proposed method's fault detection performance is much better than ANN's.

4 Conclusions

The paper introduces a simple and practical approach to improve ANN's performance in single and multiple damage identification of a girder bridge. The number of hidden layers and neurons is optimized based on a stochastic optimization process instead

Table 8 Results of damage identification for two-damage scenario

Actual damage			ANN			AGTOANN		
Elem 1	Elem 2	Level	Elem 1	Elem 2	Level	Elem 1	Elem 2	Level
3	15	9.4	5	16	9.09	3	15	9.08
5	9	24.3	5	9	25.11	5	9	24.63
1	4	18	2	4	18.67	1	4	18.32
2	10	18.3	3	10	18.45	2	10	18.29
1	9	26.5	3	11	26.50	1	9	26.27
2	10	16.6	3	10	16.72	2	10	16.61
2	8	33.7	4	7	33.87	2	8	33.69
6	12	33.3	8	11	33.22	6	12	33.43
1	16	24.5	2	14	24.81	1	16	24.27
2	3	23.3	3	4	23.60	2	3	22.99
4	5	21.6	8	12	21.22	4	5	21.63
12	16	34.8	12	16	35.97	13	16	34.93
6	13	37.6	6	12	38.07	6	13	37.64
8	14	19.2	6	13	19.54	8	14	19.23
12	13	25.6	9	13	26.10	12	13	25.66
7	16	21.8	5	12	21.92	7	16	21.71
9	11	28.8	6	10	29.12	9	11	29.03
1	4	20.9	3	5	21.36	1	4	20.98
6	15	4.2	5	11	4.27	6	15	4.10

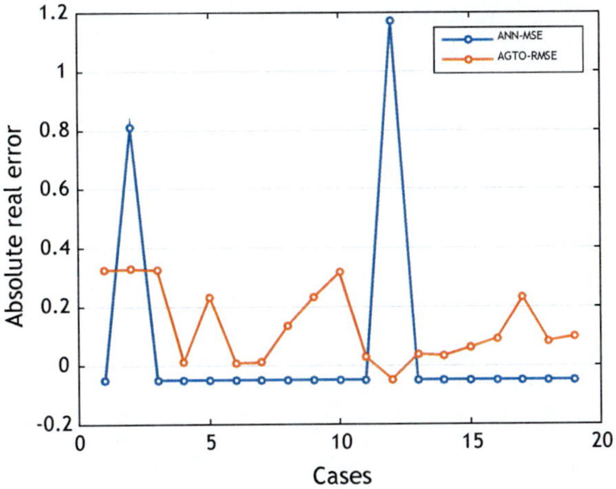

Fig. 17 Sum-up results of damage identification for the second scenario

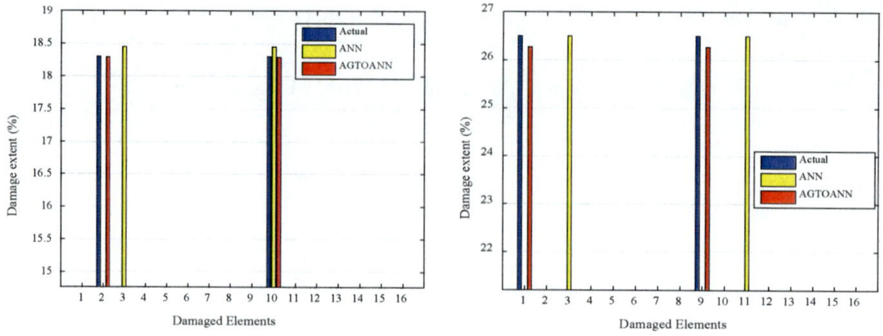

a. 18.3% at elements 2, 10

b. 26.5% at elements 1, 9

Fig. 18 Multi-damage identification using ANN and AGTOANN

of the designer's experience. This study also investigates several evaluation metrics for the regression model. The outcomes of the study can be outlined as follows:

- By identifying the best Silverback from the gorilla troop, the AGTO algorithm demonstrates its potential to improve the prediction ability of ANN. This improvement is made by optimizing the number of hidden layers and neurons. The two damage scenarios showed the effectiveness of using AGTO in network training. Especially, in the case of multi-damage identification, the proposed approach shows superiority in determining the location and extent of damage compared to ANN.
- The findings also indicate that in this current study, the two-evaluation metrics, MAE and RMSE, can achieve better results than the others.
- However, reducing computational time should be considered in further works.

References

1. Bui-Tien T, Ho LV, Quang NT (2021) A hybrid heuristic optimization algorithm PSOGSA coupled with a hybrid objective function using ECOMAC and frequency in damage detection. *J Mater Eng Struct (JMES),* 8(1), pp 31–45
2. Alkayem NF, Shen L, Asteris PG, Sokol M, Xin Z, Cao M (2022) A new self-adaptive quasi-oppositional stochastic fractal search for the inverse problem of structural damage assessment. Alex Eng J 61(3):1922–1936. https://doi.org/10.1016/j.aej.2021.06.094
3. Ferreira Gomes G, Souza Chaves JA, de Almeida FA (2021) An inverse damage location problem applied to AS-350 rotor blades using bat optimization algorithm and multiaxial vibration data. *Mech Syst Signal Process,* 145, https://doi.org/10.1016/j.ymssp.2020.106932
4. Nguyen-Ngoc L, Tran-Ngoc H, Bui-Tien T, Mai-Duc A, Magd Abdel Wahab, Nguyen Huan X, De Roeck G (2021) Damage detection in structures using Particle Swarm Optimization combined with Artificial Neural Network. *Smart Struct Syst,* 28(1), p 12, https://doi.org/10.12989/sss.2021.28.1.001

5. Ho BX, Trinh TT, Ho LV (2022) Swarm intelligence-based technique to enhance performance of ANN in structural damage detection. Transp Commun Sci J 73(1):1–15. https://doi.org/10.47869/tcsj.73.1.1
6. Ho LV, Nguyen DH, de Roeck G, Bui-Tien T, Wahab MA (2021) Damage detection in steel plates using feed-forward neural network coupled with hybrid particle swarm optimization and gravitational search algorithm. J Zhejiang Univ-SCIENCE A 22(6):467–480. https://doi.org/10.1631/jzus.A2000316
7. Zhu ZH, Ye ZF, Tang Y (2021) Nondestructive identification for gender of chicken eggs based on GA-BPNN with double hidden layers. *J Appl Poult Res,* 30(4), https://doi.org/10.1016/j.japr.2021.100203
8. Wan W, Mabu S, Shimada K, Hirasawa K, Hu J (2009) Enhancing the generalization ability of neural networks through controlling the hidden layers. Appl Soft Comput 9(1):404–414. https://doi.org/10.1016/j.asoc.2008.01.013
9. Ogunbo JN, Alagbe OA, Oladapo MI, Shin C (Jun 2020) N-hidden layer artificial neural network architecture computer code: geophysical application example. Heliyon 6(6):e04108. https://doi.org/10.1016/j.heliyon.2020.e04108
10. Tran-Ngoc H, Khatir S, Le-Xuan T, De Roeck G, Bui-Tien T, Abdel Wahab M (2020) A novel machine-learning based on the global search techniques using vectorized data for damage detection in structures. *Int J Eng Sci,* 157, https://doi.org/10.1016/j.ijengsci.2020.103376
11. Boger POBZ, Guterman H (1997) Knowledge extraction from artificial neural networks models In: IEEE International Conference on Systems, Man, and Cybernetics. Computational Cybernetics and Simulation, 4, p 6, https://doi.org/10.1109/ICSMC.1997.633051.
12. Ho LV, Trinh TT, De Roeck G, Bui-Tien T, Nguyen-Ngoc L, Abdel Wahab M (2022) An efficient stochastic-based coupled model for damage identification in plate structures, *Eng Fail Anal,* 131, https://doi.org/10.1016/j.engfailanal.2021.105866
13. Ho LV et al. (2021) A hybrid computational intelligence approach for structural damage detection using marine predator algorithm and feedforward neural networks, *Comput & Struct,* 252, https://doi.org/10.1016/j.compstruc.2021.106568.
14. Tran-Ngoc H, Khatir S, De Roeck G, Bui-Tien T, Abdel Wahab M (2019) An efficient artificial neural network for damage detection in bridges and beam-like structures by improving training parameters using cuckoo search algorithm. *Eng Struct,* 199, https://doi.org/10.1016/j.engstruct.2019.109637
15. Padil KH, Bakhary N, Abdulkareem M, Li J, Hao H (2020) Non-probabilistic method to consider uncertainties in frequency response function for vibration-based damage detection using Artificial Neural Network. *J Sound Vib,* 467, https://doi.org/10.1016/j.jsv.2019.115069
16. Khatir S, Tiachacht S, Le Thanh C, Ghandourah E, Mirjalili S, Abdel Wahab M (2021) An improved artificial neural network using arithmetic optimization algorithm for damage assessment in FGM composite plates. *Compos Struct,* 273 https://doi.org/10.1016/j.compstruct.2021.114287
17. K. M. Cort J. Willmott (2005) Advantages of the mean absolute error (MAE) over the root mean square error (RMSE) in assessing average model performance. *Clim Res,* 30(1), p 4, https://doi.org/10.3354/cr030079
18. Chai T, Draxler RR (2014) Root mean square error (RMSE) or mean absolute error (MAE)?—Arguments against avoiding RMSE in the literature. Geosci Model Dev 7(3):1247–1250. https://doi.org/10.5194/gmd-7-1247-2014
19. Karunasingha DSK (2022) Root mean square error or mean absolute error? Use their ratio as well. Inf Sci 585:609–629. https://doi.org/10.1016/j.ins.2021.11.036
20. Tran-Ngoc H, Khatir S, Ho-Khac H, De Roeck G, Bui-Tien T, Wahab MA (2021) Efficient artificial neural networks based on a hybrid metaheuristic optimization algorithm for damage detection in laminated composite structures, (in English). *Compos Struct,* 262, Apr 15 2021. https://doi.org/10.1016/j.compstruct.2020.113339.

21. Abdollahzadeh B, Gharehchopogh FS, Mirjalili S (October 2021) Artificial gorilla troops optimizer: A new nature-inspired metaheuristic algorithm for global optimization problems, (in English). Int J Intell Syst 36(10):5887–5958. https://doi.org/10.1002/int.22535
22. ANSYS, 275 Technology Drive, Canonsburg, PA 15317, Release 17.2

Numerical Analyses for Evaluation of Factor of Safety Distribution Map

Alfrendo Satyanaga, Sung-Woo Moon, Martin Wijaya, Sonny Irawan, and Jong Kim

Abstract Climate change has resulted in the formation of rainfall behaviour that is aberrant from historical trends. Consequently, this leads to escalation of the dangers of slope failures due to rainfall. Factor of safety distribution map is a common tool used to record the possibility of slope failure, by indicating the factor of safety (FoS) of slopes in an area. In this study, FoS distribution map was developed for the Bukit Timah Granite formation in Singapore utilizing Transient Rainfall Infiltration and Grid-based Regional Slope-Stability Model (TRIGRS). Several slopes were analysed to evaluate the accuracy of the FoS distribution map. The analyses were performed via 2D numerical analyses involving seepage and slope stability analyses on the investigated slopes that have experienced slope failure in the past. The results of the analyses indicated that the FoS from the map were lower as compared to those from 2D numerical analyses. Potential sources of the discrepancy could arise from the usage of infinite slope model as well as 1D vertical infiltration in the TRIGRS software. To minimize the errors associated with the 1D infiltration and infinite slope model, correction factors were proposed in this study.

Keywords Factor of safety · Bukit Timah Granite · Seepage · Slope stability

1 Introduction

Soils above the groundwater table have been observed to exhibit negative pore-water pressure which is also termed soil suction and is further defined as the energy required for extracting a unit volume of water from the soil [1]. During rainfall events, water infiltrates the soil and reduces matric suction, leading to reduced shear strength of soils and inducing slope failures [2]. On the other hand, if rainfall intensity is higher

A. Satyanaga (✉) · S.-W. Moon · S. Irawan · J. Kim
Nazarbayev University, Nur-Sultan, Kazakhstan
e-mail: alfrendo.satyanaga@nu.edu.kz

M. Wijaya
Parahyangan Catholic University, Bandung, Indonesia

M. Abdel Wahab (ed.), *Proceedings of the 5th International Conference on Numerical Modelling in Engineering*, Lecture Notes in Civil Engineering 311,
https://doi.org/10.1007/978-981-19-8429-7_3

than the permeability, a fraction of the flux will not be able to infiltrate the soil, resulting in surface runoff [3].

Finite element method is commonly used to analyse complex problems in the present day [4]. It involves the usage of primarily triangular and quadrilateral meshes and nodes to be iterated for the solution of boundary-value problems. This provides a more precise and efficient analysis method compared to the conventional hand calculation [5]. For unsaturated soil conditions with transient soil–water characteristic curve (SWCC) and permeability functions, the complex nature of the analysis warrants the usage of finite element method since hand calculation would be time-consuming and challenging [6]. SEEP/W and SLOPE/W from GeoStudio are the commercially available softwares used for 2D finite element method analysis of seepage and slope stability, respectively [7–9].

For seepage analysis on SEEP/W, requisite input parameters include slope geometries, groundwater table, soil properties and boundary conditions. Slope geometries and the groundwater table level can be drawn accordingly for analysis. Soil properties comprise soil densities, SWCC and permeability function. These properties can be imported or estimated based on commonly used equations [10–12]. Lastly, boundary conditions are the defining components of boundary-value finite element analysis. These entail the application of a flux or hydraulic head condition at boundaries [13]. The seepage analysis is used to generate pore-water pressure distribution which would be imported into SLOPE/W for slope stability analysis [14].

Pre-requisite input parameters for slope stability analysis in SLOPE/W include unsaturated shear strength, soil densities and entry and exit planes of slip surfaces. In addition, specification of a grid of potential centre of rotation is required for the calculation of spatial and temporal variations. These are used to generate slip surfaces for the identification of the critical slip surface and minimum FoS of the slope [15]. Although non-circular slip surfaces have been shown to provide lower FoS, circular slip surfaces have been traditionally employed as they enable a methodical approach to studying multiple trial slip surfaces with lower computational effort and more reliable results [16].

The objective of this study is to perform seepage and slope stability analyses for evaluation of the existing FoS distribution map within residual soil from Bukit Timah Granite in Singapore. The research works in this study include 2D finite element seepage and limit equilibrium slope stability analyses for selected residual soil slopes within Bukit Timah Granite in Singapore.

2 Methodology

Seepage analyses were carried out with the SEEP/W software. Equation 1 below illustrates the partial differential equation on water infiltration [1]. This allows for the development of the pore-water pressure contours, in 2D seepage analyses.

$$\frac{\partial}{\partial x}\left(-k_{wx}\frac{\partial h_w}{\partial x}\right)+\frac{\partial}{\partial y}\left(-k_{wy}\frac{\partial h_w}{\partial y}\right)+q=m_2^w\gamma_w\frac{\partial h_w}{\partial t} \quad (1)$$

in which:

m_2^w	slope of the soil–water characteristic curve.
γ_w	unit weight of water.
h_w	total head or hydraulic head.
t	time.
k_{wx}, k_{wy}	permeability coefficients.
q	flux assigned externally at boundaries.

After pore-water pressure distributions were generated from analyses using SEEP/W, these distributions at every time interval were then imported into the SLOPE/W software where slope stability analysis was conducted using the limit equilibrium method. The slope stability analyses in this project are implemented using Bishop's simplified method of slices. This method considers no shear forces between slices and the sole forces between slices forces act normal (horizontally) to each slice [17]. Although this method does not provide for horizontal force equilibrium, vertical force and moment equilibrium are satisfied and studies have shown that it produces results that are accurate [18, 19]. Although more rigorous and complex methods exist, it was found that for shallow slip surfaces, Bishop's simplified method produces similar results [20, 21].

The FoS distribution map consists of contours of the FoS of a region. It is often used in urban planning or hazard management. It is developed using factors such as the angle of the slope, geological formation, rainfall and usage of the land. This is of growing importance as land use gradually expands to sloping terrain. It is compounded by extreme weather patterns which involve more intense rainfall [22, 23]. As a result, soil infiltration is increased and shear strength is lowered, causing landslides.

TRIGRS is a software applied to model the locations and timing of landslides, in particular, shallow slope failures. Richard's 1D vertical infiltration equation is used for modelling storm water to derive the FoS of each zone [16]. The model assumes a slope of infinite length, homogeneous soil characteristics and that the slip surface is parallel to the ground. The FoS results are then used to create a map of FoS contours.

3 Numerical Analyses

The FoS distribution map was extracted from [16] (Fig. 1). It was used to obtain the FoS values for comparison with the FoS results from the SLOPE/W. The evaluation of the FoS distribution map was carried outing using five residual soil slopes from Bukit Timah Granite, that have experienced landslides in the past for the analysis. Past records of slope failure were obtained from Singapore Land Authority. The selected slopes chosen are (i) Sembawang Hills Drive, (ii) Sembawang Road, (iii)

Yishun Ave 2, (iv) Marsiling Road and (v) Ang Mo Kio Ave 4. The locations of the slopes and boreholes are shown in Fig. 1. The slope angles and heights for each slope are shown in Table 1. The digital elevation map was then used to gather data on the slope height and length to model in GeoStudio 2012. As Yishun Ave 2 and Sembawang Road are located close to each other, data from the same borehole was used for modelling both slopes. Additionally, Ang Mo Kio Ave 4 and Sembawang Hill Drive share the same slope geometry.

The list of SWCC parameters [24] used to best fit the SWCC of residual soils from each slope is presented in Table 2. The permeability function used in the analysis was derived using the statistical method. The saturated permeability (ks) values are also shown in Table 2.

An example of a slope modelled is shown in Fig. 2. The boundary was modelled at least three times the slope height away from the slope, to limit its influence on

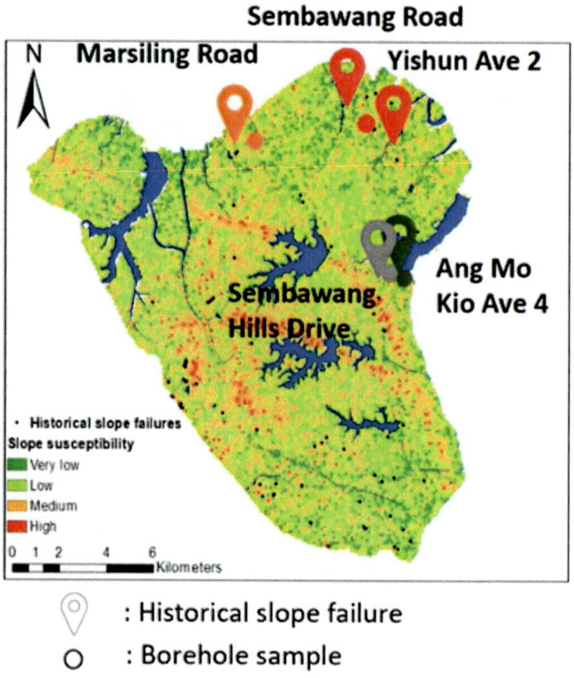

Fig. 1 Location of slopes failures and boreholes, in the Bukit Timah Granite SSM

Table 1 Slope geometries

Slope	Angle (°)	Height (m)
Sembawang Hill Drive, Ang Mo Kio Ave 4	41.2	7
Sembawang Road	40.6	6
Yishun Ave 2	56.3	9
Marsiling Road	36.9	6

Table 2 SWCC fitting parameters and k_s of residual soils from each slope

Slope	a	n	m	θ_s	k_s
Sembawang Hill Drive	10.2	1.0	3.2	0.32	4.7e−5
Ang Mo Kio Ave 4	1544	1.1	2.6	0.6	4.7e−8
Sembawang Road, Yishun Ave 2	11.3	1.4	0.9	0.37	2.8e−5
Marsiling Road	101	0.5	1.3	0.6	6.0e−6

seepage at the slope. The boundary conditions applied are (a) no flux and seepage at the bottom, (b) 22 mm/hr rainfall at the top and (c) the head at the lateral boundaries were applied with respect to the water table. A 22 mm/hr unit flux was applied in accordance with the largest rainfall recorded in Singapore [2]. Prior to transient seepage analyses under rainfall loading, preliminary analyses were performed for each slope to generate a natural-looking shape of groundwater table as illustrated in Fig. 3 The water table is within 2–3 m below the toe of the slope. Additionally, suction was limited to −75 kPa [2] to avoid unreasonably large suction values on surfaces of the slope. A 22 mm/hr flux was applied for 1 day. Following that, rainfall was stopped, and the pore-water pressure contours were recorded for at least 48 h while drying.

The stability analyses were performed by importing pore-water pressure distributions from all time steps in seepage analyses to Slope/W [25]. Table 3 shows the shear strength of soil used in the stability analyses. As no tension cracks were observed on the slopes, it was not incorporated into the analyses.

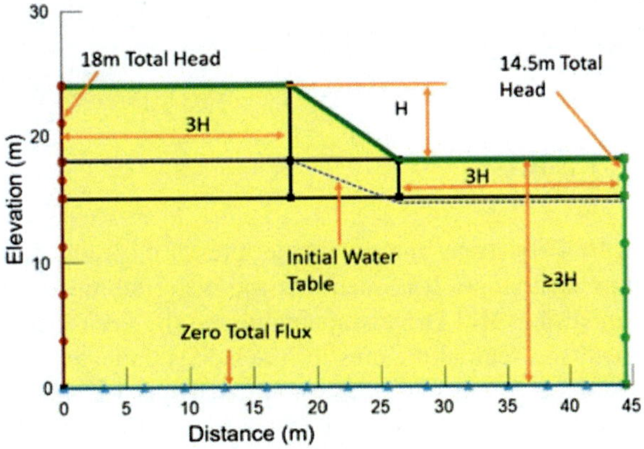

Fig. 2 Numerical model of slope at Marsiling Road

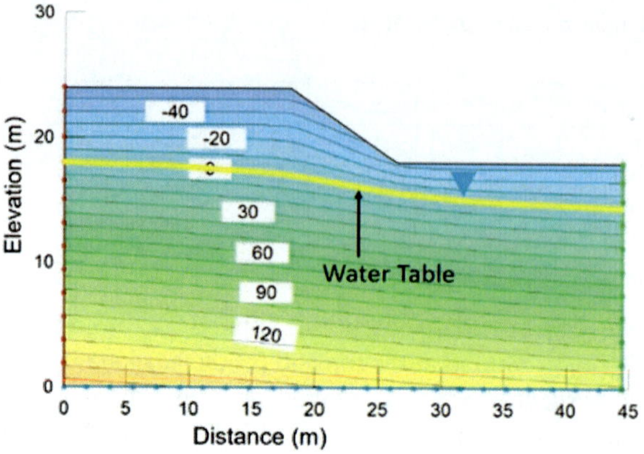

Fig. 3 Water table after 24 h of no rainfall

Table 3 Shear strength of residual soils from each slope

Slope	Effective cohesion, c' (kPa)	Effective friction angle, ϕ' (°)	Unsaturated shear strength angle, ϕ^b (°)	Bulk density, ρ (kN/m^3)
Sembawang Hill Drive	10.2	1.0	3.2	0.32
Ang Mo Kio Ave 4	1544	1.1	2.6	0.6
Sembawang Road, Yishun Ave 2	11.3	1.4	0.9	0.37
Marsiling Road	101	0.5	1.3	0.6

4 Results

Figure 4 illustrates the pore-water pressure contours after 24 h of 22 mm/hr rainfall from each slope. It is seen that the water table rose to different levels, where the Sembawang Hill Drive slope exhibited the least increase. Also, the band of unsaturated soil seen at the crest of the slopes indicates that the water table and wetting front have yet to meet. On the other hand, the wetting front and water table have merged at the slope toe in all slopes except the Sembawang Hill Drive slope. This is due to the lateral seepage of water downward the slope.

Figure 5 shows the FoS contours, the critical slip surface and its respective FoS, at the end of the 24 h of 22 mm/hr rainfall. The critical slip surfaces are shallow and run through the wetting fronts of the slope. This is a reasonable observation, since reduced suction occurs at the wetting front, causing lower shear strength.

FoS was also recorded against time as shown in Fig. 6. All slopes showed recovery of FoS immediately after the end of rainfall, with the exception of AMKA, where

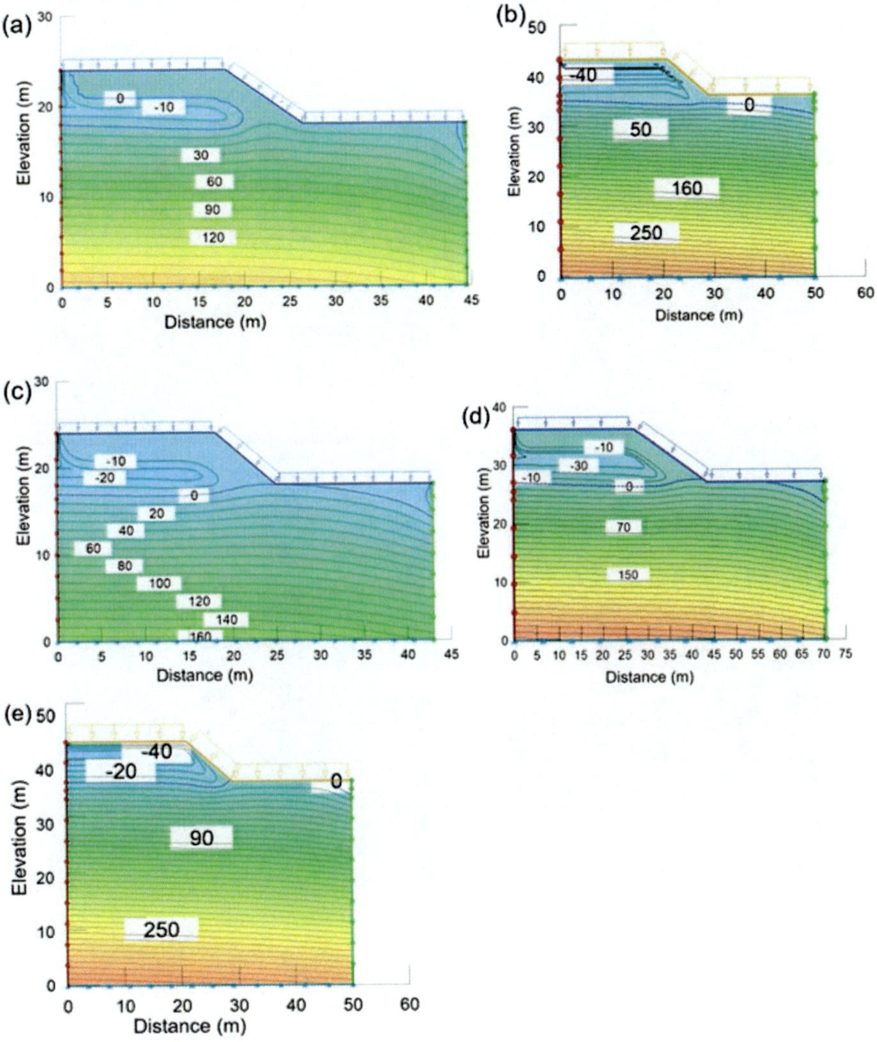

Fig. 4 Pore-water pressure contours of **a** Marsiling Road slope; **b** Sembawang Hill Drive slope; **c** Sembawang Road; **d** Yishun Avenue 2; **e** Ang Mo Kio Ave 4

FoS reduced slowly until 84 h after the rainfall has stopped. This is due to the low ks value in Ang Mo Kio Ave 4. As a result, although the rainfall has ended, the water within the soil continued to permeate slowly, causing the FoS to continue reducing. Table 4 shows the original and minimum FoS, as well as the percentage reduction in FoS. The most change in FoS was seen in Sembawang Hill Drive, while the least change was seen in Ang Mo Kio Ave 4. This is due to the small c' and large ϕ^b values in Sembawang Hill Drive, causing a large change in FoS in response to the reduction in suction.

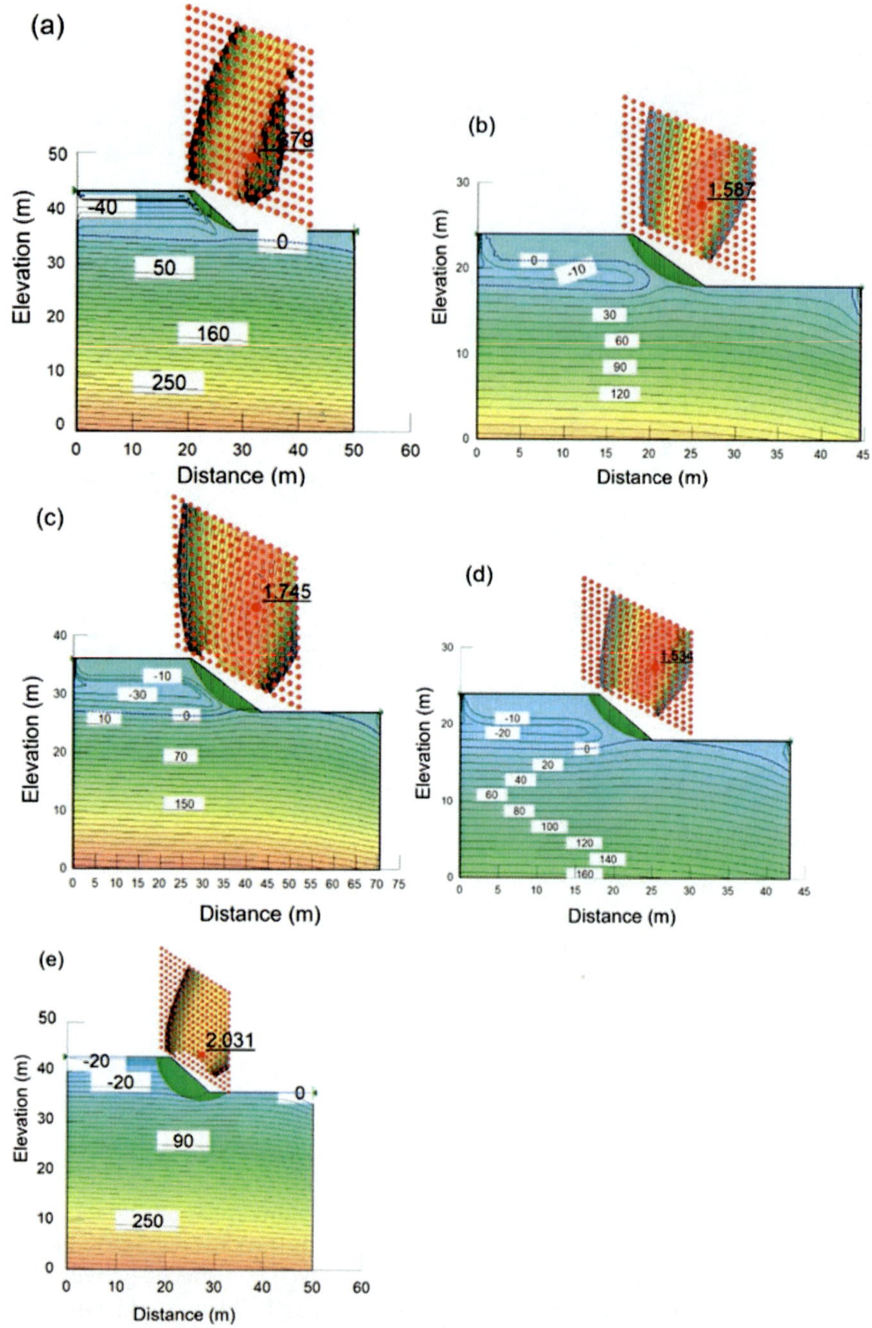

Fig. 5 Critical Slip surface associated with minimum FoS of **a** Marsiling Road slope; **b** Sembawang Hill Drive slope; **c** Sembawang Road; **d** Yishun Avenue 2; **e** Ang Mo Kio Ave 4

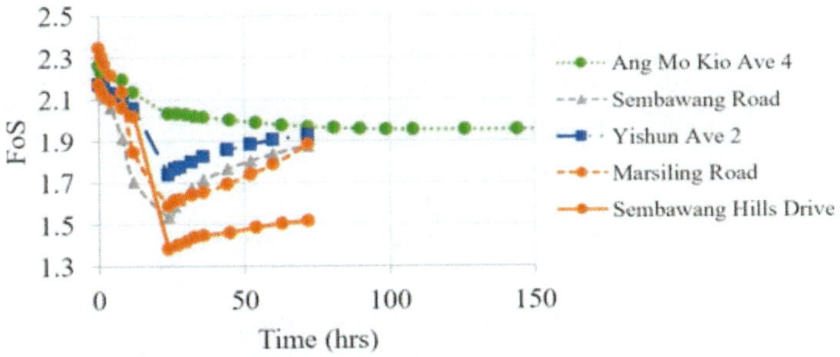

Fig. 6 FoS variations against time

Table 4 Original, minimum and percentage reduction of FoS

Slope	Original FoS	Minimum FoS	% Δ FoS
Sembawang Hill Drive	2.35	1.6	32.2
Ang Mo Kio Ave 4	2.17	1.38	36.5
Sembawang Road	2.17	1.53	29.2
Yishun Ave 2	2.17	1.74	20.0
Marsiling Road	2.26	1.95	13.7

5 Discussions

The FoS values from the FoS distribution map and the minimum FoS from the 2D analyses are compiled in Table 5. As Sembawang Hill Drive and Ang Mo Kio Ave 4 slopes have the same geometry, the lowest FoS from the two slopes from the 2D analyses were used for a more conservative comparison. It can be observed that the FoS distribution map produces FoS results lower than that from the 2D analyses. The most significant difference was observed in Yishun Ave 2 slope.

Table 5 FoS from FoS distribution map and 2D analyses

Slope	FoS		% Δ FoS
	FoS distribution map	2D analysis	
Sembawang Hill Drive	1.10	1.38	0.28
Sembawang Road	1.34	1.53	0.19
Yishun Ave 2	1.09	1.74	0.65
Marsiling Road	1.28	1.59	0.31

Table 6 $\frac{d}{l}$ values of the slopes

Slope	d/l
Sembawang Hill Drive	0.216
Sembawang Road	0.307
Yishun Ave 2	0.475
Marsiling Road	0.251

This is due to the usage of TRIGRS, which applies the: (i) 1D infiltration and (ii) infinite slope model. Since 1D infiltration limits water seepage laterally in the soil matrix, it causes pore-water pressure to build up, lowering suction, and thus, the soil shear strength and FoS decrease. Yishun Ave 4 slope has the steepest slope angle and it has the highest permeability. As a result, more lateral flow was observed within the soil matrix in this slope, hence resulting in the most deviation of the FoS obtained from FoS distribution map and 2D analysis. In addition, the infinite slope model does not satisfy moment equilibrium and causes larger driving forces which leads to smaller FoS in FoS distribution map. The error associated with the infinite slope model was found to be related to the failure plane depth (d) and length of the slope (l). This error increases with d/l, and when d/l is above 0.1, the error associated with infinite slope model is at least 20% (Huang et al. 2015). Table 6 shows the d/l values of the slopes. It can be seen that all slopes have d/l of more than 0.1 and YA has the largest d/l value. This indicates at least a 20% error in all slopes, and the largest error in YA, corresponding to the largest deviation in FoS from the FoS distribution map and 2D analyses.

6 Conclusions

The FoS distribution map yielded lower FoS values compared to 2D cross-sectional analyses. The difference is caused by the fact that TRIGRS software applies infinite slope model and 1D infiltration. To adjust for the errors associated with these two models, correction factors may be incorporated. To improve the map's accuracy, more borehole data can be collected for usage in the FoS distribution map, especially since residual soils are known to exhibit high variability.

Acknowledgements The authors appreciate the financial support from Nazarbayev University, Social Policy Grant and Grants 11022021CRP1512. Any opinions, findings and conclusions or recommendations expressed in this material are those of the author(s) and do not necessarily reflect the views of Nazarbayev University.

References

1. Fredlund DG, Rahardjo H (1993) Soil mechanics for unsaturated soils. John Wiley & Sons
2. Satyanaga A, Rahardjo H (2020) Role of unsaturated soil properties in the development of slope susceptibility map. Geotech Eng
3. Satyanaga A, Moon S-W, Kim JR (2022) Stability analyses of dual porosity soil slope. Geomech Eng 28(1):77–87
4. Tran-Ngoc H, Khatir S, Le-Xuan T, De Roeck G, Bui-Tien T, Wahab MA (2021) Finite element model updating of a multispan bridge with a hybrid metaheuristic search algorithm using experimental data from wireless triaxial sensors. Eng Comput. https://doi.org/10.1007/s00366-021-01307-9
5. Wang S, Wahab MA (2021) A numerical study on the effect of variable wear coefficient on fretting wear characteristics. Materials 14:1840. https://doi.org/10.3390/ma14081840
6. Rahardjo H, Satyanaga A, Leong EC, Wang J-Y (2014) Comprehensive instrumentation for real time monitoring of flux boundary conditions in slope. Procedia Earth Planet Sci 9:23–43
7. Kristo C, Rahardjo H, Satyanaga A (2019) Effect of hysteresis on the stability of residual soil slope. Int Soil Water Conserv Res 7:226–238
8. Kim Y, Rahardjo H, Nistor MM, Satyanaga A, Leong EC, Sham AWL (2022) Assessment of critical rainfall scenarios for slope stability analyses based on historical rainfall records in Singapore. Environ Earth Sci 81(39):1–13
9. Chan YEC, Ng QL, Satyanaga A, Rahardjo H (2020) Regional stability and adaptation measures against rainfall-induced slope failures. Environ Geotech 1–16
10. Zhai Q, Rahardjo H, Satyanaga A, Dai G, Zhuang Y (2020) Framework to estimate the soil-water characteristic curve for soils with different void ratios. Bull Eng Geol Environ
11. Zhai Q, Rahardjo H, Satyanaga A, Dai G (2020) Estimation of the soil-water characteristic curve from the grain-size distribution of coarse-grained soils. Eng Geol 267:105502
12. Satyanaga A, Zhai Q, Rahardjo H (2017) Estimation of unimodal water characteristic curve for gap-graded soil. Soils Found 57(5):789–801. https://doi.org/10.1016/j.sandf.2017.08.009
13. Rahardjo H, Satyanaga A, Harnas FR, Leong EC (2016) Use of dual capillary barrier as cover system for a sanitary landfill in Singapore. Ind Geotech J 46(3):228–238
14. Rahardjo H, Kim Y, Satyanaga A (2019) Role of unsaturated soil mechanics in geotechnical engineering. Int J Geo-Eng 10(8)
15. Satyanaga A, Wijaya M, Zhai Q, Moon S-W, Pu J, Kim JR (2021) Stability and consolidation of sediment tailings incorporating unsaturated soil mechanics. Fluids 6:423
16. Ip CY, Rahardjo H, Satyanaga A (2021) Three-dimensional slope stability analysis incorporating unsaturated soil properties in Singapore. Georisk, Assess Manage Risk Eng Syst Geohazards 15(2):98–112
17. Bishop AW (1955) The use of the slip circle in the stability analysis of slopes. Geotechnique 5:7–17
18. Skempton AW, Brown J (1961) A landslide in boulder clay at Selset, Yorkshire. Geotechnique 11:280–293
19. Spencer E (1967) A method of analysis of the stability of embankments assuming parallel inter-slice forces. Geotechnique 17:11–26
20. Ching R, Fredlund D (1984) Quantitative comparisons of limit equilibrium methods of slices. Methods 73:379
21. Fredlund DG, Krahn J (1977) Comparison of slope stability methods of analysis. Can Geotech J 14:429–439

22. Camici S, Ciabatta L, Brocca L, Moramarco T (2014) Impact of climate change on flood frequency using different climate models and downscaling approaches. J Hydrol Eng 19
23. Ip CY, Rahardjo H, Satyanaga A (2019) Spatial variations of air-entry value for residual soils in Singapore. Catena 174:259–268
24. Fredlund DG, Xing A (1994) Equations for the soil-water characteristic curve. Can Geotech J 31:533–546
25. Rahardjo H, Satyanaga A (2019) Sensing and monitoring for assessment of rainfall-induced slope failures in residual soil. Geotechn Eng 172(GE6)

A Hybrid Optimization Algorithm for Structural Health Monitoring

H. H. Tran-Ngoc, T. Le-Xuan, N. Hoang-Thanh, L. Dao-Dac, T. Bui-Tien, and M. Abdel Wahab

Abstract Over the last decades, optimization algorithms, especially Particle Swarm Optimization (PSO) have been applied successfully in many fields. PSO contains a great capacity of global search that helps it converge faster in the first steps. However, in the local search stage, the convergence speed of PSO reduces significantly, especially when the elements are close to the global best. This reduces the effectiveness and increases the computational cost of PSO. In this work, we propose combining Firefly Algorithm (FA) and PSO to deal with optimization problems. This algorithm can employ the global search capacity of PSO and the local search capacity of FA. The proposed method is applied for Structural Health Monitoring (SHM) for a large-scale truss bridge. The obtained result shows that the proposed method surpasses traditional PSO and FA in terms of accuracy.

Keywords Particle Swarm Optimization · Firefly Algorithm · Structural Health Monitoring

1 Introduction

During the service life, engineering structures are easily subjected to damages caused by a wide range of reasons such as natural impacts (storms, floods, earthquakes, and

H. H. Tran-Ngoc (✉) · M. Abdel Wahab
Department of Electrical Energy, Metals, Mechanical Constructions, and Systems, Faculty of Engineering and Architecture, Ghent University, 9000 Gent, Belgium
e-mail: ngochoa@utc.edu.vn

H. H. Tran-Ngoc · T. Le-Xuan · T. Bui-Tien
Department of Bridge and Tunnel Engineering, Faculty of Civil Engineering, University of Transport and Communications, Hanoi, Vietnam

N. Hoang-Thanh
Department of Science and Technology, Ministry of Transport, Hanoi, Vietnam

L. Dao-Dac
Specialized Department, University of Transport Technology, Thai Nguyen Campus, Thai Nguyen, Vietnam

M. Abdel Wahab (ed.), *Proceedings of the 5th International Conference on Numerical Modelling in Engineering*, Lecture Notes in Civil Engineering 311,
https://doi.org/10.1007/978-981-19-8429-7_4

so on) and human impacts (overload, accident, and so on). These incidents not only severely affect the economy but also threaten the safety of traffic participants. Therefore, in recent decades, Structural Health Monitoring (SHM) has received special attention from scientific communities reported in the literature [1]. For instance, Samir et al. [2] proposed a new robust flexibility index to detect damages (level and location) of three complex structures consisting of a 52-bar space truss, a 52-bar planar truss, and a 37-bar planar truss. The stiffness of the truss joint of a large-scale truss bridge was identified in the work of [3]. In this research, the author employed One-Dimensional Convolutional Neural Network (1DCNN) to analyze the reduction of stiffness.

With the striking development of science and technology, in recent decades, numerous optimization algorithms have been proposed and widely applied to a wide range of fields. Among the proposed algorithms, PSO has proven effective to deal with SHM problems. Nguyen-Ngoc et al. [4] proposed a novel method combining PSO and Artificial Neural Network (ANN) to detect damage of beam structures. In this work, both numerical model and measurement were applied. The obtained results showed that the proposed method possibly identified damages in the considered structures with high accuracy. Tran et al. [5] applied PSO to detect damages in a steel beam structure. An improved PSO combined with ANN was used for damage assessment of a simply supported reinforced concrete beam bridge and a free-free beam in the laboratory [6].

However, it is admitted that PSO still has disadvantages such as local search capacity. This reduces the effectiveness and increases the computational cost of PSO. To remedy the shortcomings of PSO, in this paper, we propose a hybrid algorithm that combines PSO and FA, termed HFAPSO. This algorithm can employ the global search capacity of PSO and the local search capacity of FA.

The efficiency of the proposed algorithm is evaluated by monitoring the health of a large-scale truss bridge. To compare with the proposed algorithm, the traditional PSO and FA are also employed.

2 Methodology

2.1 Traditional Particle Swarm Optimization

In 1995, Kennedy et al. [7] developed PSO based on stochastic optimization, inspired by the social behavior of animals such as a flock of birds, a school of fish, and so on when they look for food. Each element moves in the search space and shares information with other ones, which significantly increases the chance to obtain the food. The working principle of PSO is based on two main equations. The first one (1) is to update the position of each element.

$$X^{t+1} = X^{t} + V^{t+1} \tag{1}$$

where $\boldsymbol{X}^t$, and $\boldsymbol{X}^{t+1}$are the current position of the element at the t and t +1 iterations, respectively. The second Eq. (2) is to calculate the velocity of each element in the next iteration.

$$\boldsymbol{V}^{t+1} = w'\boldsymbol{V}^t + C'r'(\boldsymbol{P}^t - \boldsymbol{X}^t) + C''r''(\boldsymbol{G}^t - \boldsymbol{X}^t) \tag{2}$$

where r' and r'' are random values between [0, 1]; C' and C'' are the learning parameters; and w' is the weight parameter. Each element owns a locally optimal solution ($\boldsymbol{P}^t$), whereas $\boldsymbol{G}^t$ is the global optimal solution of tth iteration. Elements are randomly initialized and move through the search space to seek new solutions. These solutions are compared with each other, and the best solution is selected.

2.2 *Firefly Algorithm*

FA is a metaheuristic algorithm developed by Yang [8]. This algorithm imitates the behavior of fireflies. Fireflies use light to perform a variety of activities such as communicating, hunting, and threatening predators. According to the inverse square law, the light intensity (I) at a distance r from the light source (l_s) can be calculated using the Eq. (3).

$$I = l_s/r^2 \tag{3}$$

Equation (3) can be formulated based on Gaussian (see Eq. (4)):

$$F = F_0\exp(-\gamma r^2) \tag{4}$$

where F is the attractiveness of a firefly at a distance (r) and F_0 is the attractiveness, when r= 0; Light absorption coefficient (γ) ∈ [0, ∞].

Assuming that i and j are two fireflies and their positions are X_i (x_i, y_i) and X_j (x_j, y_j), respectively, the distance (RE) between two fireflies is calculated based on Euclidean using Eq. (5):

$$RE = ||X_i - X_j|| = \sqrt{\left(x_i - x_j\right)^2 - \left(y_i - y_j\right)^2} \tag{5}$$

The new position of the less bright firefly (X_i) moving towards the more attractive firefly j is updated with Eq. (6):

$$X_i^{t+1} = X_i^t + F_0\exp(-\gamma RE^2)\left(X_j^t - X_i^t\right) + a\varepsilon_i \tag{6}$$

where ε_i is the vector of random variables and a is a random parameter (a) ∈ [0, 1].

The working principle of FA is depicted in Fig. 1.

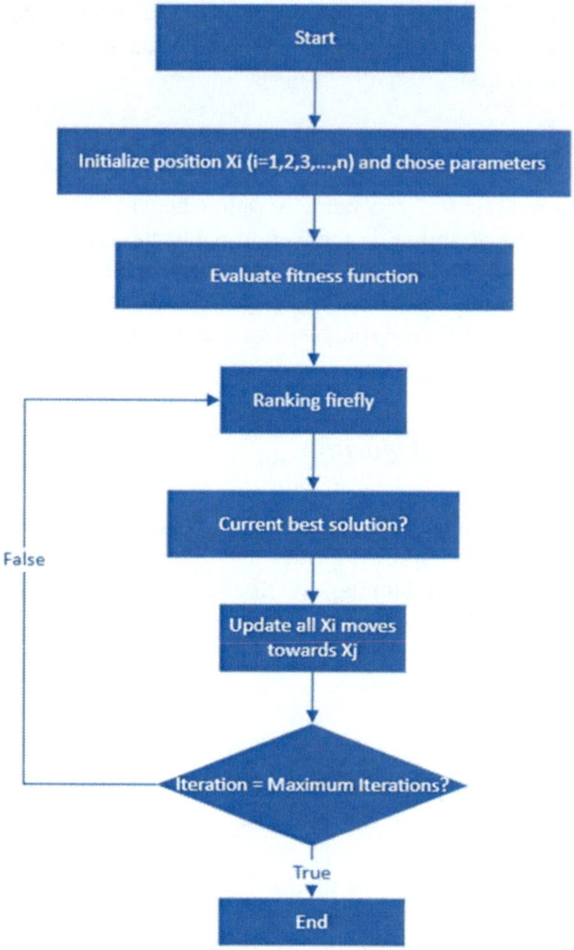

Fig. 1 Firefly Algorithm flowchart

2.3 HFAPSO

In this paper, an optimization algorithm (HFAPSO) that combines the advantages of two algorithms PSO (global search ability) and FA (local search ability) is proposed. First, w' (weight parameter) is replaced by the linear decreasing inertia weight (w) using Eq. (7).

$$w = w_{start} - ((w_{start} - w_{end})/\text{Maximum Iterations}) \times \text{Iteration} \quad (7)$$

The best values $\boldsymbol{P}^t$ and $\boldsymbol{G}^t$ are evaluated. If $\boldsymbol{P}^t < \boldsymbol{G}^t$, HFAPSO is applied for searching local solutions (Eqs. (8) and (9)). Conversely, if $\boldsymbol{P}^t > \boldsymbol{G}^t$, PSO is applied (Eqs. (1) and (2)).

Update new position.

$$X_i^{t+1} = X_i^t + F_0 \exp(-\gamma RE^2)\left(X_i^t - \mathbf{G}^t\right) + a\varepsilon_i \tag{8}$$

Calculate new velocity (Fig. 2).

$$V_i^{t+1} = X_i^{t+1} - X_i^t \tag{9}$$

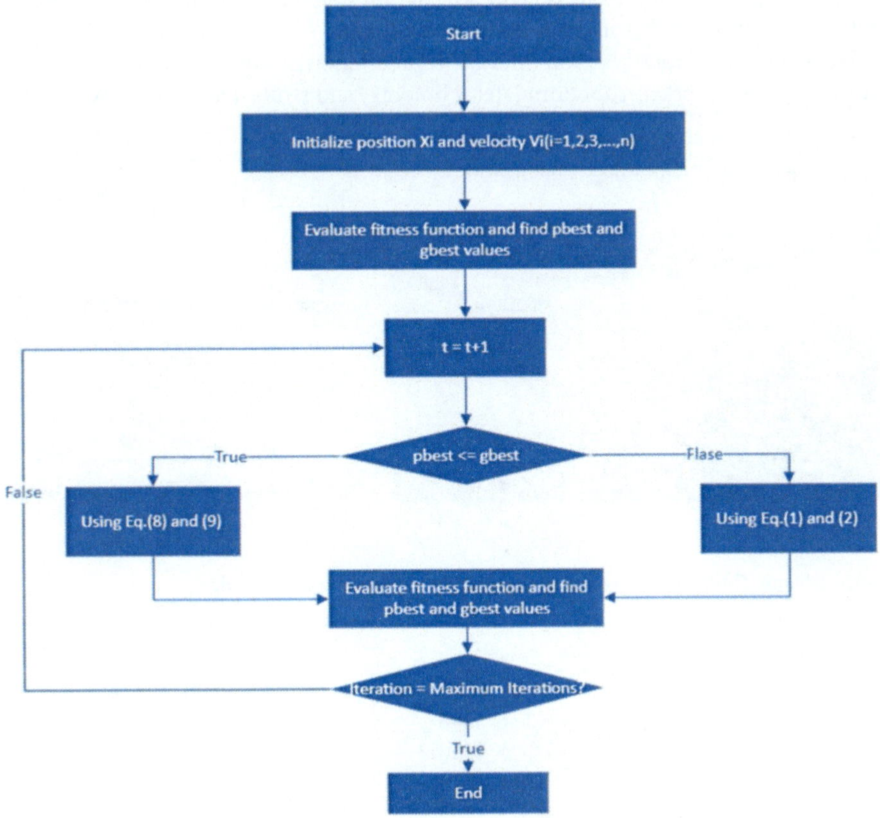

Fig. 2 HFAPSO flowchart

3 Application of HFAPSO for Damage Detection of a Large-Scale Bridge

3.1 Description of the Bridge

To consider the effectiveness of the proposed method, in this section, a large-scale bridge is employed. Chuong Duong Bridge (Fig. 3) is a steel truss bridge, crossing the Red River, on National Highway 1 at km 170 + 200, Hanoi city, Vietnam. This bridge includes 11 simple spans with almost equal lengths (90 m) connecting the center of Hoan Kiem district with Long Bien district.

Truss members comprise top chords, vertical chords, bottom chords, diagonal chords, top lateral bracing, bottom lateral bracing, and struts depicted in Fig. 4 [9].

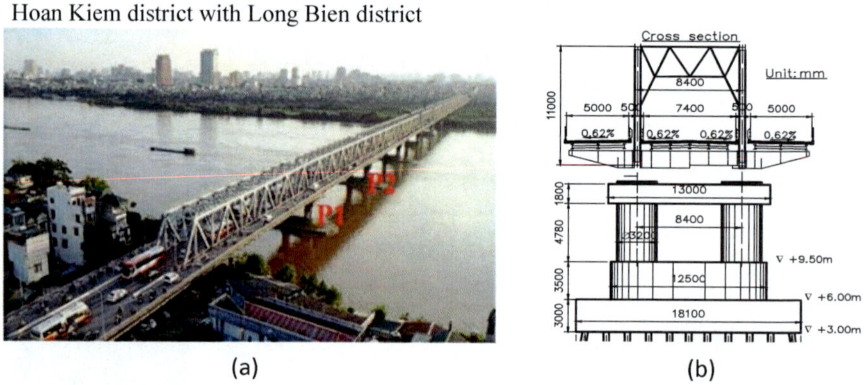

Fig. 3 Chuong Duong bridge; **a** general layout; **b** cross section of the bridge

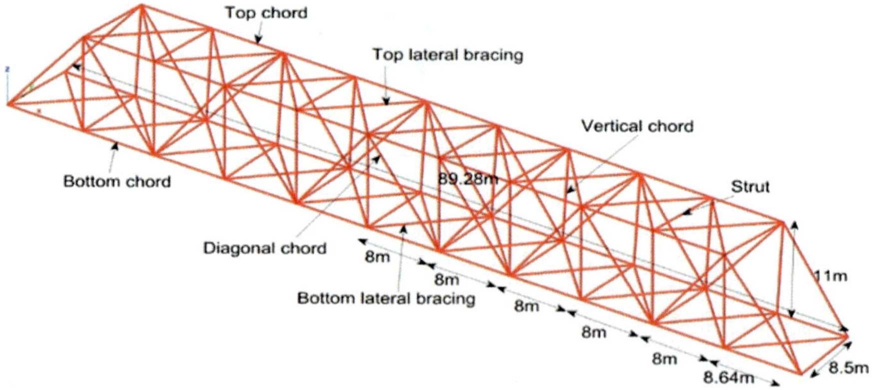

Fig. 4 Truss members

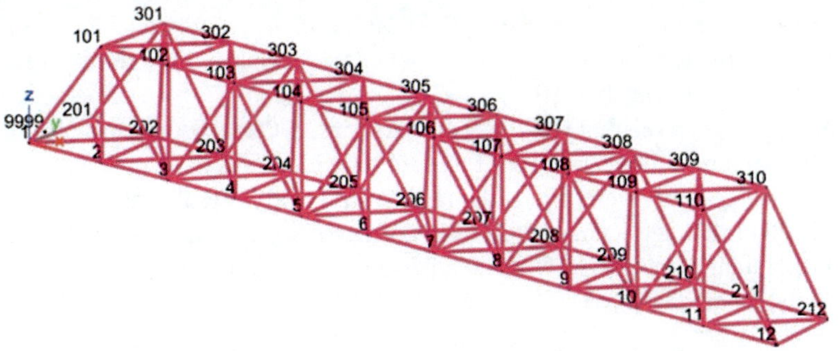

Fig. 5 FEM of Chuong Duong bridge

3.2 Finite Element Model

In this paper, the first span ($P1$-$P2$) is chosen to assess the effectiveness of the proposed approach. The Finite Element Model (FEM) of this span is built using the Stabil toolbox of MATLAB [10]. The model consists of 45 nodes and 145 elements. Each element has 6 degrees of freedom (DOFs) including translational and rotational displacements in the X, Y, Z-axis (see Fig. 5).

While a pin bearing is put on $P1$ pier, a roller is used for $P2$ Pier. Pin bearing only permits rotational displacements, whereas roller one allows both translational and rotational displacements. Non-structural components consisting of parapets, maintenance paths, and so on are taken into account as added mass. The X-axis corresponds to the longitudinal direction, the Y-axis coincides with the horizontal axis of the bridge, and the Z-axis is in the vertical direction.

3.3 Damage Detection

In this section, the proposed method (HFAPSO) is employed to detect damages to the Chuong Duong bridge. To compare with HFAPSO, traditional PSO and FA are also used. Some main parameters of PSO, FA, and HFAPSO used in this paper are described in Table 1.

Table 1 Parameters of algorithms

Algorithms	Parameters	Values
PSO	w'	0.9
	C', C''	(c_1', c_2') 2, 2
FA	α	0.5
	β	0.2
	γ	1

The process of looking for the optimal solution is completed based on two conditions: (1) the maximum iteration is 100 or the deviation between the real and calculated value is lower than 10^{-7}.

From Fig. 6, it can see that HFAPSO outperforms PSO and FA in terms of convergence level. Figure 7 shows that HFAPSO not only identifies damage location accurately but also possibly determines damage level exactly. Traditional FA and PSO only can detect damage locations. In terms of detecting damage level, some errors occur. For instance, the level of damage predicted by FA and PSO is 40 and 44%, respectively, whereas the real damage level is 50%.

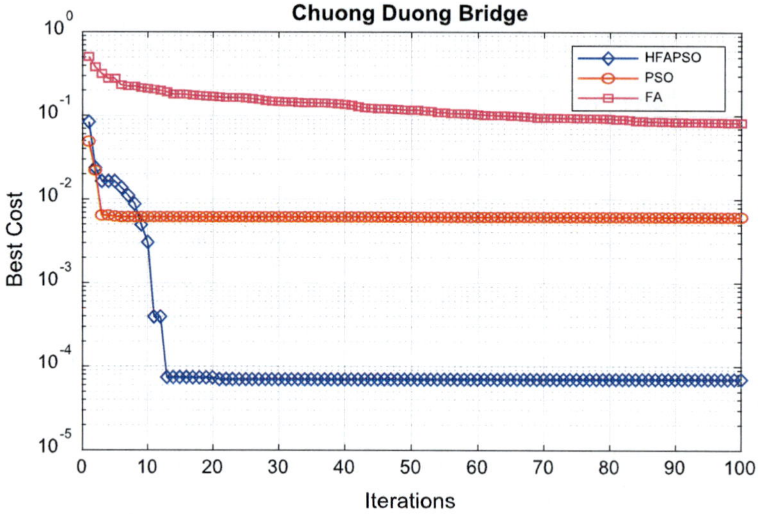

Fig. 6 Convergence

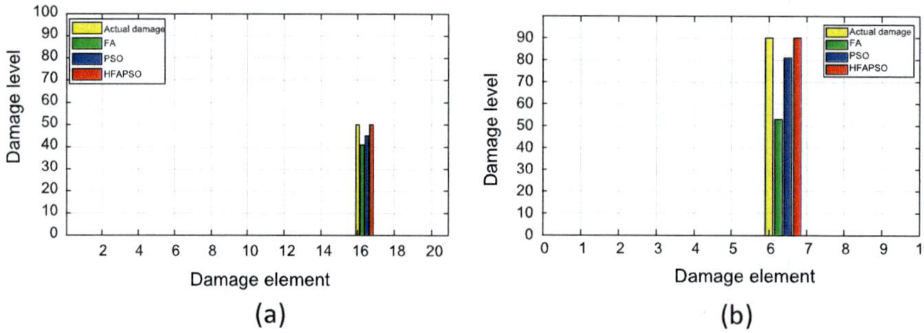

Fig. 7 Damage detection **a** 50% damage, **b** 90% damage

4 Conclusions

This paper focuses on detecting damages in the Chuong Duong bridge using optimization algorithms. To remedy the shortcomings of traditional PSO and FA, a hybrid HFAPSO is proposed. This algorithm can employ the global search capacity of PSO and the local search capacity of FA. For comparison purposes, traditional PSO and FA are also employed. Based on obtained results, some remarks are given below.

- All considered algorithms possibly identify the damage location in the Chuong Duong bridge exactly.
- HFAPSO outperforms PSO and FA in terms of accuracy when detecting damage levels.
- In this paper, the proposed method is only employed to detect damages to the numerical model. Further research should employ HFAPSO to deal with real problems.

Acknowledgements The authors acknowledge the financial support of VLIR-UOS TEAM Project, VN2018TEA479A103, "Damage assessment tools for Structural Health Monitoring of Vietnamese infrastructures," funded by the Flemish Government. This work is also funded by Vingroup and supported by Innovation Foundation (VINIF) under project code VINIF.2021.DA00192. Moreover, the first author needs to acknowledge the financial support from the University of Transport and Communications (UTC) under the project research "T2022-CT-017".

References

1. Dang HV, Tran-Ngoc H, Nguyen TV, Bui-Tien T, De Roeck G, Nguyen HX (2020) Data-driven structural health monitoring using feature fusion and hybrid deep learning. IEEE Trans Autom Sci Eng 18(4):2087–2103
2. Khatir S, Tiachacht S, Le Thanh C, Tran-Ngoc H, Mirjalili S, Wahab MA (2021) A new robust flexibility index for structural damage identification and quantification. Eng Fail Anal 129:105714
3. Dang HV, Raza M, Tran-Ngoc H, Bui-Tien T, Nguyen HX (2021) Connection stiffness reduction analysis in steel bridge via deep CNN and modal experimental data. Struct Eng Mech Int J 77(4):495–508
4. Nguyen-Ngoc L, Tran H, Bui-Tien T, Mai-Duc A, Abdel Wahab M, Nguyen XH, De Roeck G (2021) Damage detection in structures using particle swarm optimization combined with artificial neural network. Smart Struct Syst 28(1):1–12
5. Tran NH, Bui TT (2019) Damage detection in a steel beam structure using particle swarm optimization and experimentally measured results. Sci J Transp 9:3–9
6. Tran-Ngoc H, Khatir S, Le-Xuan T, Tran-Viet H, De Roeck G, Bui-Tien T, Wahab MA (2022) Damage assessment in structures using artificial neural network working and a hybrid stochastic optimization. Sci Rep 12(1):1–12
7. Kennedy J, Eberhart R (1995) Particle swarm optimization. In: Proceedings of ICNN'95-international conference on neural networks, vol 4. IEEE, pp 1942–1948
8. Yang XS, He X (2013) Firefly algorithm: recent advances and applications. arXiv:1308.3898

9. Tran-Ngoc H, Khatir S, Le-Xuan T, De Roeck G, Bui-Tien T, Wahab MA (2020) A novel machine-learning based on the global search techniques using vectorized data for damage detection in structures. Int J Eng Sci 157:103376
10. François S, Schevenels M, Dooms D, Jansen M, Wambacq J, Lombaert G, Degrande G, De Roeck G (2021) Stabil: an educational Matlab toolbox for static and dynamic structural analysis. Comput Appl Eng Educ 29(5):1372–1389

Transient Analysis of Heat Transfer in a Trunk Under a Forest Fire Influence

Eusébio Conceição, João Gomes, Mª Manuela Lúcio, Domingos Viegas, and Mª Teresa Viegas

Abstract This article describes a numerical study, in a transient regime, on the heat transfer of heat and mass in a trunk under a forest fire influence. The geometries of the trunk and fire front were obtained by mesh generation. These meshes are used to calculate the view factors that will be used later in the estimation of the radiative exchanges between the trunk and the fire front. The thermal response model of the trunk is based on energy balance integral and differential equations. Heat exchanges are defined by conduction inside the trunk, by convection between the trunk bark and the surrounding ambience and by radiation between the trunk bark and the surroundings, including the fire front. In this study, it is assumed that the fire front propagates with a constant value of fire spread rate of 0.1 m/s and that the flame temperature is 500ºC. The temperature distribution evolution inside the trunk as well as in its bark was obtained for a variable and random wind speed with an average value equal to 12 m/s and a standard deviation equal to 8 m/s. In general, there are slight fluctuations in temperature depending on fluctuations in wind speed.

Keywords Forest fire · Numerical evaluation · Trunk · Thermal behaviour · Wind speed

---

E. Conceição (✉) · M. M. Lúcio
Universidade do Algarve, 8005-139 Faro, Portugal
e-mail: econcei@ualg.pt

J. Gomes
CINTAL, 8005-139 Faro, Portugal
e-mail: jgomes@ualg.pt

D. Viegas · M. T. Viegas
Universidade de Coimbra, 3030 Coimbra, Portugal
e-mail: xavier.viegas@dem.uc.pt

M. T. Viegas
e-mail: maria.viegas@dem.uc.pt

M. Abdel Wahab (ed.), *Proceedings of the 5th International Conference on Numerical Modelling in Engineering*, Lecture Notes in Civil Engineering 311,
https://doi.org/10.1007/978-981-19-8429-7_5

1 Introduction

Humidity, topography, wind and temperature are among the main influences on how fire spreads in forests [1]. The higher the temperature, the lower the humidity and the lower the humidity, the drier the air, therefore dry fuel ignites and burns more easily. The slope of the landscape is also important: fires burn much faster uphill than downhill [2]. A fire originates radiation and convection phenomena that preheat the unburned fuel ahead of the fire front more effectively upslope than down. Fire spread is significantly affected by wind speed [3, 4]. Wind can bias fire actions by pushing moist air away from fuels, drying them out quickly. It transports burning sparks that have been raised by convection air, which will start new fires far beyond the perimeter. It bends the convection column, which causes preheating of unburned fuels ahead of the fire front. It continually feeds the fire with oxygen. Strong winds have a pressing major result on the wildfire spread rate when encountering dry fuels [5]. The environmental wind fluctuates in both direction and speed, with increased air instability implying greater fluctuations [3]. Works on wildfire spread models [6], simulations of spread and behaviour of real fires [7] and fire spreads with variable wind strengths [8] present relevant aspects about the behaviour of forest fires when under the effect of fluctuations in wind speed.

The numerical model utilized here to simulate the thermal behaviour of the trunk was based on a numerical model utilized to simulate the thermal behaviour of a human body [9, 10], both considering a similar way of using equations of energy balance in boundary conditions. These energy balance equations define processes of heat transfer as follows: by conduction within the trunk; by convection between the trunk bark and the surrounding ambience; and by radiation existing between bark of the trunk bark and the surroundings (fire front included). View factors are used in the evaluation of the radiative exchanges between the pine trunk bark and the surroundings. The view factors evaluation sub-model is defined in a similar way to the ones used to obtain the buildings' thermal response [11, 12] or the thermal response of vehicle cabins [13].

The objective of this numerical study is to build up a model that simulates the transient thermal response of a tree under the effect of a wildfire, in which the environmental variable wind has a random speed. Thus, the tree is typified by its trunk and the wildfire is typified by a front fire with a constant spread rate. The transient thermal response will be evaluated by the temperature distribution evolution within and on the trunk surface (bark).

2 Numerical Model

The thermal behaviour of the trunk is computed by a numerical model, which is constituted by differential energy equations and it uses a grid generated to represent the trunk. The trunk consists of bark and cambium. The numerical model ponders

Fig. 1 Mesh generation of the trunk

energy balance equations at the boundary between the trunk and the surrounds considering the following thermal phenomena: heat conduction within the trunk, heat convection (natural, forced and mixed) between the bark and the environment air and radiative heat exchanges between the bark and the surfaces of the surrounding bodies, namely, the sky, the fuel surface and the fire front.

The following hypotheses are considered in the establishment of the energy balance equations:

- The heat flux is processed considering two dimensions;
- The temperature of the air surrounding the trunk is uniform, with a value identical to the temperature of the ambient air, and will increase with the approach of the fire front;
- Recourse to heat transfer coefficients by convection determined for isothermal surfaces;
- Fire actions around the trunk are neglected.

Numerical simulation results depend on the generated mesh. Here, it is used a mesh adapted to the trunk geometry developed from the finite differences method. The numerical mesh generation considers two spaces: a physical one and a computational one. With this method, the physical domain is transformed into the computational plan by the use of two elliptic partial differential equations, of Poisson's type. Figure 1 presents the mesh generation utilized in the numerical simulation.

3 Methodology

The numerical simulation results to be analysed are those acquired in a transient regime, which refer to the evolution of the temperature distribution inside and on the surface (bark) of the trunk. In this simulation, it was considered that the wind speed is random and it has an average value of 12 m/s with a standard deviation of 8 m/s (Fig. 2).

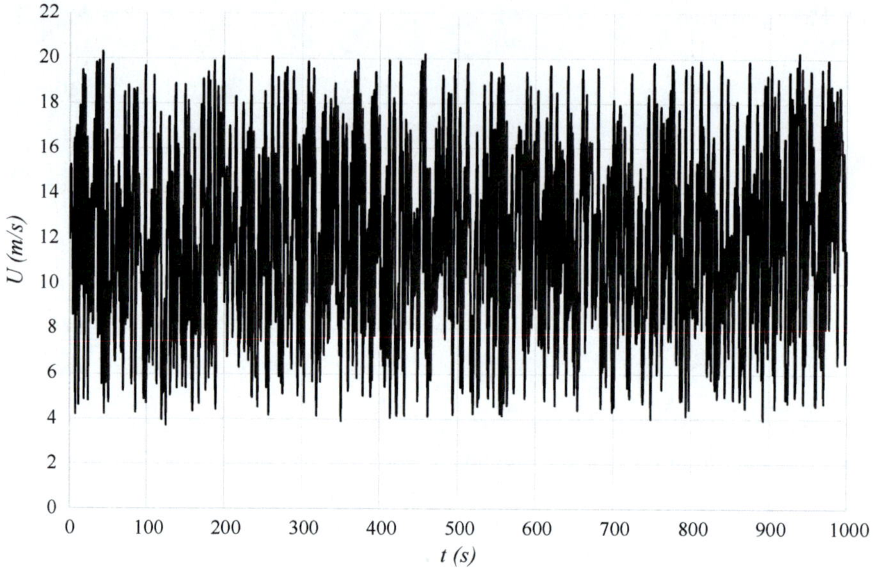

Fig. 2 Evolution of wind speed

During the simulation, it is assumed that the fire front propagates from an initial position located 4 m upstream of the trunk, with a constant spread rate equal to 0.01 m/s. The height and outer diameter of the trunk are equal to 0.5 m and 0.4 m, respectively. The characteristics of the fire front are as follows: 2 m high; 1 m wide; 45° tilt; and an average flame temperature of 500 °C. The representation of fire propagation towards the trunk is shown in Fig. 3. In the vicinity of the trunk, the following environmental conditions are verified: air temperature of 20°C; 50% relative humidity. The temperature distribution was acquired at points located on the trunk bark and radially inside the trunk. Points located equidistantly on the trunk bark in a longitudinal plane at a height of 0.5 m, totaling 33, are designated by P. Points located radially inside the trunk, totaling 20, are designated by Q. A cross section of the trunk showing where the 33 P-points on its bark and the 20 Q-points on its interior are located is presented in Fig. 4.

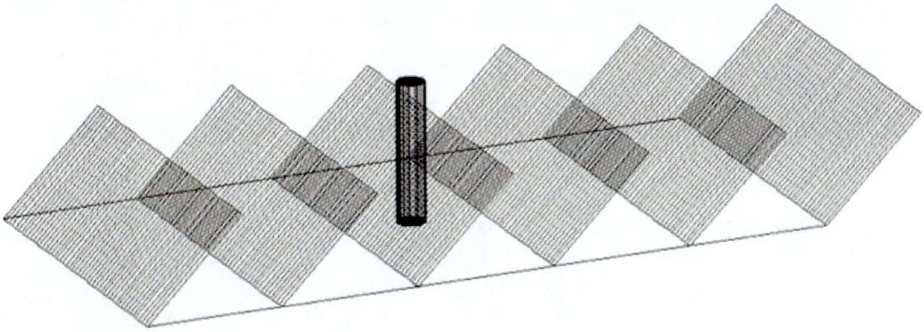

Fig. 3 Fire propagation towards the trunk

Fig. 4 Cross section of the trunk showing where the 33 P-points on its bark and the 20 Q-points on its interior are located

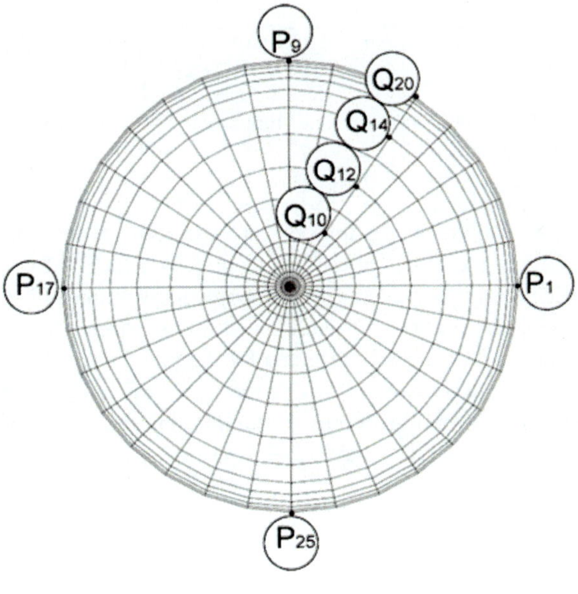

4 Results and Discussion

The evolution of the temperature at the 33-P points located on the trunk bark is presented in Fig. 5. The results show that, during the progression of the fire front, the fire initially affects the bark on the upstream side of the trunk. Considering the trunk bark, the highest temperatures are reached at points located on the upstream side due to the slope of the flame. It is found that fluctuations in wind speed cause fluctuations in the calculated temperatures of the trunk bark.

The temperature evolution at the 20-Q points placed on the interior of the trunk, containing point P9 on the upstream side and point P25 on the downstream side (see Fig. 4), is presented in Fig. 6.

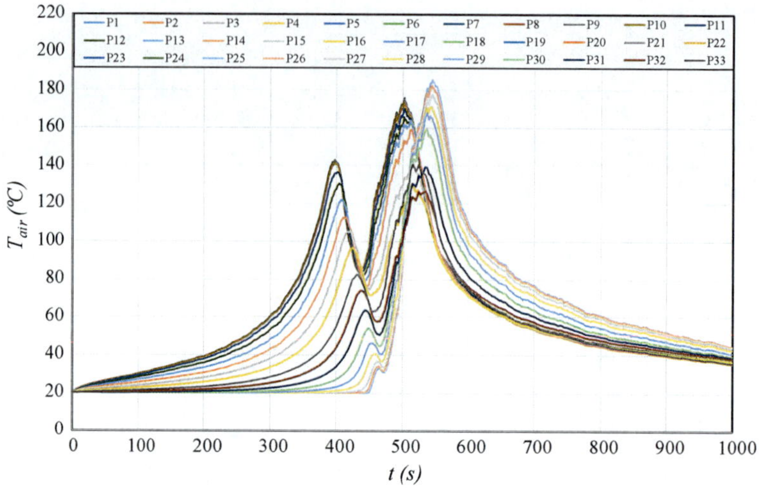

Fig. 5 Evolution of the temperature at the 33-P points located on the trunk bark considering a random wind speed

As shown by the temperature variation in relation to the ambient temperature of 20°C, the effect of the fire front only affects the four outermost rings. The most affected ring is the outermost one (given by point Q19, as point Q20 refers to the trunk bark). Fluctuations in wind speed do not affect the temperature of the trunk rings, which are essentially affected by the passage of the fire front and the temperature of its flame. In general, inside the trunk, the temperatures calculated at the points of the rings turned upstream are higher than those calculated at the points of the rings turned downstream due to the slope angle of the flame in the direction of progression of the fire front.

5 Conclusions

This article presented a numerical study carried out in a forest fire environment in order to obtain the thermal response of a pine tree (here represented only by its trunk) subjected to the action of this type of intense fire. The analysis of the results obtained focused on the temperature field in the bark and inside the pine trunk obtained in a transient regime. This study was done assuming a random wind speed and a fire front progressing towards the trunk with a constant spread rate.

Taking into account the considered values of the average wind speed and its standard deviation, the wind speed fluctuations are only reflected in fluctuations of the temperature obtained in the trunk bark. The temperatures inside the pine trunk are not conditioned by the randomness of the wind speed. Points on the upstream

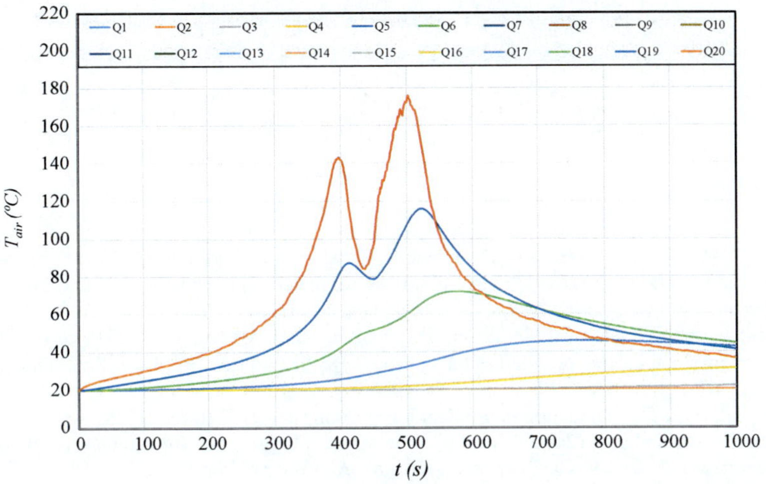

a)

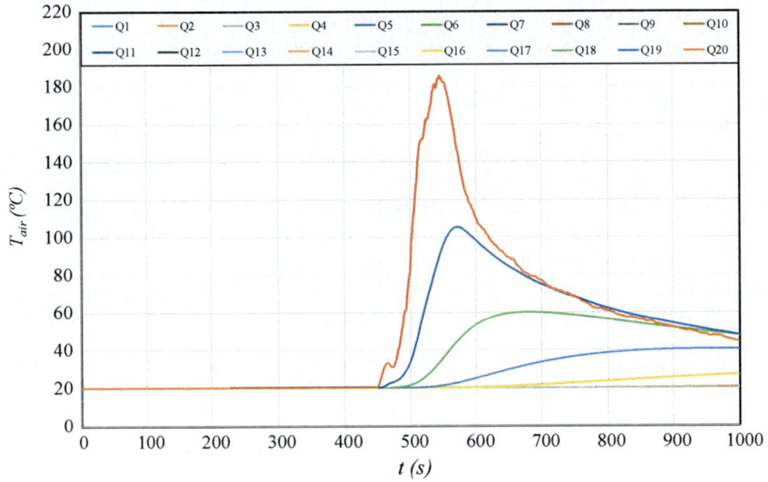

b)

Fig. 6 Evolution of the temperature at the 20-Q points placed on the interior of the trunk containing the bark points: **a** P9; **b** P25. Note that point Q20 corresponds to the point located on the pine trunk bark

side of the trunk exhibit higher temperatures than those on its downstream side due to the flame slope towards the movement of the fire front.

Following this work and as a proposal for future work, we will numerically study the influence of the angle of inclination of a forest fire front on the tree thermal response.

Acknowledgements The authors would like to acknowledge the support of the project reference PCIF/MPG/0108/2017, funded by the Portuguese Foundation of Science and Technology (FCT).

References

1. Tošić I, Mladjan D, Gavrilov M, Živanović S, Radaković M, Putniković S, Petrović P, Mistridželović I, Marković S (2019) Potential influence of meteorological variables on forest fire risk in Serbia during the period 2000–2017. Open Geosci 11:414–425
2. Eftekharian E, Ghodrat M, He Y, Ong R, Kwok K, Zhao M, Samali B (2019) Investigation of terrain slope effects on wind enhancement by a line source fire. Case Stud Therm Eng 14:100467
3. Beer T (1991) The interaction of wind and fire. Bound-Layer Meteorol 54:287–308
4. Cruz M, Alexander M (2019) The 10% wind speed rule of thumb for estimating a wildfire's forward rate of spread in forests and shrublands. Ann For Sci 76:44
5. Cruz M, Alexander M, Fernandes P, Kilinc M, Sil A (2020) Evaluating the 10% wind speed rule of thumb for estimating a wildfire's forward rate of spread against an extensive independent set of observations. Environ Model Softw 13:104818
6. Nelson R (2002) An effective wind speed for models of fire spread. Int J Wildland Fire 11:153–161
7. Jahdi R, Darvishsefat A, Etemad V, Mostafavi M (2014) Wind effect on wildfire and simulation of its spread (Case study: Siahkal Forest in Northern Iran). J Agric Sci Technol 16:1109–1121
8. Song H, Lee S (2017) Effects of wind and tree density on forest fire patterns in a mixed-tree species forest. For Sci Technol 13(1):9–16
9. Conceição E, Rosa S, Custódio A, Andrade R, Meira M, Lúcio M (2010) Study of airflow around occupants seated in desks equipped with upper and lower air terminal devices for slightly warm environments. HVAC&R Res 16(4):401–412
10. Conceição E, Lúcio M (2016) Numerical simulation of the application of solar radiant systems, internal airflow and occupants' presence in the improvement of comfort in winter conditions. Buildings 6(3):38
11. Conceição E, Nunes A, Gomes J, Lúcio M (2010) Application of a school building thermal response numerical model in the evolution of the adaptive thermal comfort level in the Mediterranean environment. Int J Vent 9(3):287–304
12. Conceição E, Lúcio M (2010) Numerical study of the influence of opaque external trees with pyramidal shape in the thermal behaviour of a school building in summer conditions. Indoor Built Environ 19:657–667
13. Conceição E, Silva M, André J, Viegas D (2000) Thermal behaviour simulation of the passenger compartment of vehicles. Int J Veh Des 24(4):372–387

Design of an Auditorium Equipped with an Attached Solar Greenhouse Used to Improve Indoor Environmental Conditions

Eusébio Conceição, João Gomes, Mª Inês Conceição, Mª Manuela Lúcio, and Hazim Awbi

Abstract This article proposes the design of an auditorium equipped with an attached solar greenhouse utilized to provide better conditions of comfort to the occupants during the winter season. The three-dimensional design of the auditorium and greenhouse is generated using geometric equations expressed in cylindrical coordinates. The thermal environmental conditions and air quality verified inside the auditorium are obtained by a research software, that simulates, in transient conditions, the air quality and the thermal behaviour of buildings, called Building Thermal Response (BTR). The thermal comfort within the auditorium is estimated by the Predicted Mean Vote (PMV) index. The air quality inside the auditorium is assessed by the concentration of carbon dioxide (CO_2) produced in the breathing process of its occupants. The numerical model implemented in BTR uses integral equations of energy and mass balance to represent the radiative, convective, conductive and evaporative phenomena associated with the thermal behaviour of the building and its occupants. The virtual auditorium and greenhouse are made up of 759 opaque and 25 transparent surfaces, respectively. It is assumed that the auditorium is occupied by 168 people. A variable ventilation process was implemented throughout the day. The evolutions of PMV and CO_2 obtained during the occupation cycle show that it is possible to ensure levels of thermal comfort and indoor air quality most of the time.

E. Conceição (✉) · M. M. Lúcio
Universidade do Algarve, 8005-139 Faro, Portugal
e-mail: econcei@ualg.pt

J. Gomes
CINTAL, 8005-139 Faro, Portugal
e-mail: jgomes@ualg.pt

M. I. Conceição
Instituto Superior Técnico, 1049-001 Lisboa, Portugal
e-mail: ines.conceicao@tecnico.ulisboa.pt

H. Awbi
University of Reading, Reading RG6 6AW, UK
e-mail: h.b.awbi@reading.ac.uk

M. Abdel Wahab (ed.), *Proceedings of the 5th International Conference on Numerical Modelling in Engineering*, Lecture Notes in Civil Engineering 311,
https://doi.org/10.1007/978-981-19-8429-7_6

Keywords Indoor air quality · Passive solar greenhouse · Thermal comfort · Three-dimensional design

1 Introduction

In the northern hemisphere, solar greenhouses attached to the south side of habitations can be used to reduce their heating requirement [1]. These greenhouses can be very efficient in improving the thermal performance of the adjacent compartments [2, 3], mainly in winter season, by contributing positively to the heating of these compartments without the use of additional active heating systems [4, 5]. Therefore, it is proposed, in this work, to numerically evaluate the environmental thermal conditions provided by a greenhouse attached to an auditorium.

The design of the auditorium and its attached solar greenhouse was generated three-dimensionally by a numerical model based on geometric equations defined in cylindrical coordinates (i.e. through angle, radius and z-coordinates). A similar methodology was applied to obtain the geometries of buildings [6] and the human body (occupant) [7]. The mesh generated in this way on the surfaces serves to calculate the solar radiation absorbed, transmitted and incident both in interior and exterior spaces, and the radiative exchanges in each space [8].

The use of specific software to analyse the thermal behaviour of buildings is essential to calculate the temperature distribution and establish energy gains [9, 10], among other issues. In this work, a research software is used to estimate the thermal response of the auditorium and the attached solar greenhouse, called Building Thermal Response (BTR). BTR is based on energy and mass balance integral equations [11]. Radiative, conductive, convective and evaporative phenomena are taken into account by energy balance integral equations utilized to compute the temperature distribution. Mass balance integral equations, utilized to evaluate the contaminants and water vapor distribution, are defined according to the convection, conduction and absorption/desorption phenomena. BTR is also made up of sub-models intended for calculating solar radiation, establishing energy and mass convection coefficients as well as radiative properties of glass, calculating air flow rate, etc. The solution of the system of equations thus constituted is obtained through the Runge–Kutta–Felberg method with error control.

Thermal comfort (TC) is commonly assessed by the Predicted Mean Vote (PMV) index [12]. This index depends on four indoor environmental parameters (air velocity, air temperature, mean radiant temperature (MRT) and relative humidity) and two personal parameters (activity level and clothing insulation level). International standards, such as ISO 7730 [13], use the PMV index to classify indoor environments, in function of their TC level, according to the following categories: A, for $-0.2 \leq \text{PMV} \leq 0.2$; B, for $-0.5 \leq \text{PMV} \leq 0.5$; C, for $-0.7 \leq \text{PMV} \leq 0.7$.

Indoor air quality (IAQ) is commonly estimated by the concentration of carbon dioxide (CO_2) [14–16]. International standard ASHRAE 62.1 [17] set a CO_2 concentration of 1800 mg/m^3 (1000 ppm) as the acceptable limit to obtain adequate IAQ conditions for occupants. However, the Portuguese standard establishes an acceptable limit that is less strict than the international standard, namely a CO_2 concentration of 2250 mg/m^3 (1250 ppm) [18].

The objective of this numerical work is to apply the BTR software to analyse the thermal behaviour of an auditorium with an attached solar greenhouse as well as the IAQ provided by the implemented ventilation process. In this sense, the TC and the IAQ will be computed using the PMV index and the CO_2 concentration, respectively.

2 Models and Materials

In this work two numerical models are used: one for the mesh generation of the auditorium and its attached solar greenhouse, called Building Geometry Design (BGD); another to analyse the thermal response of the auditorium and its attached solar greenhouse, called BTR.

BGD uses geometric equations in cylindrical coordinates (defined by the angle, radius and z-coordinates) to develop the geometry of the auditorium and attached solar greenhouse. The mesh generation and the system of the integral equations of mass and energy balance are then obtained from the developed geometry. The mesh generation is utilized to calculate the view factors and the MRT. The above-mentioned system of equations is developed both for each of the surfaces and for the air within each of the spaces. Opaque surfaces consist of several layers, while interior and transparent surfaces consist of just one layer.

Using previously developed mass and energy balance equations, BTR determines the distribution of temperatures, water vapour mass and contaminant concentration, the occupants' TC level and the IAQ level [11]. The temperature distribution is calculated on all bodies (opaque, transparent and interior) as well as on the air inside the spaces. The occupants' TC level is estimated by the average value of PMV index calculated inside the auditorium [12, 13]. The IAQ is evaluated by the average value of the CO_2 concentration calculated inside the auditorium [17, 18]. It should be noted that the IAQ mainly depends on the ventilation process implemented and the air renewal rate used.

In the numerical simulations carried out in this work, the virtual auditorium (Figs. 1 and 2a) consists of 759 opaque surfaces and the attached solar greenhouse (Figs. 1 and 2b) consists of 25 transparent surfaces. When meshing each surface is subdivided into at least 100 infinitesimal areas.

During the occupancy cycle, the ventilation process proceeds as follows: outside air is introduced into the solar greenhouse, where it is previously heated naturally by incident solar radiation before being subsequently channeled into the auditorium where it is exhausted to the outside. These values are shown as a percentage of the airflow rate value recommended by the standards [17, 18] for the number of occupants (168 in this study) present in the ventilated space. The ventilation process used corresponds to the one that, after several simulations, was chosen as the one

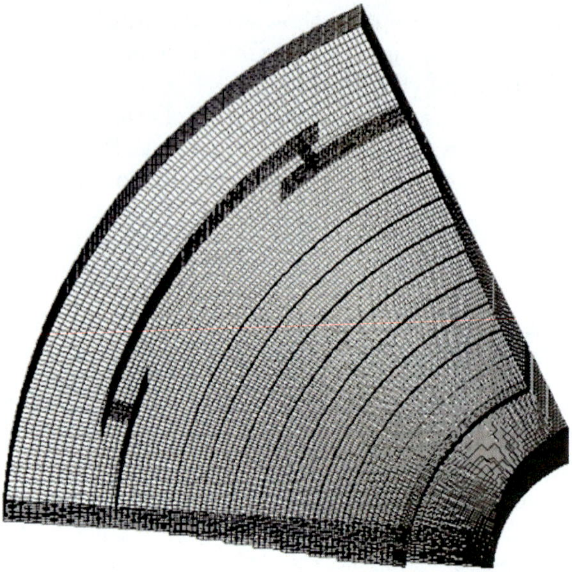

Fig. 1 Mesh generation of the virtual auditorium and its attached greenhouse

that ensures the best compromise between acceptable levels of thermal comfort and IAQ throughout the occupancy cycle.

Numerical simulations were performed for six consecutive days in order to allow the results obtained to stabilize their values. In the next section, the results presented correspond to the sixth day of the numerical simulation.

3 Results and Discussion

3.1 Indoor Air Quality

The IAQ is evaluated by the daily evolution of the CO_2 concentration whose results obtained over the course of a day can be seen in Fig. 3. During the morning, the IAQ can be considered acceptable according to the Portuguese standard [18] for CO_2 concentration values around the acceptable limit of 2250 mg/m^3; however, considering the international standard [17], the IAQ is unacceptable but for CO_2 concentration values that slightly exceed the acceptable limit of 1800 mg/m^3. During the afternoon, the IAQ is acceptable for CO_2 concentration values within the acceptable limit [17, 18].

Fig. 2 Mesh generation: **a** greenhouse; **b** auditorium

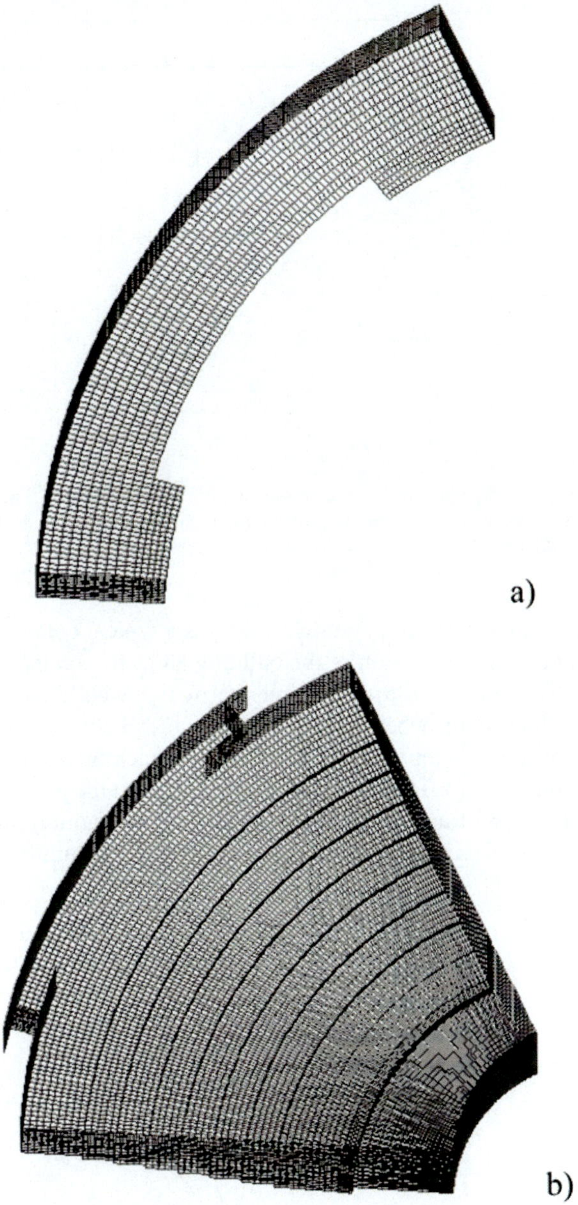

3.2 Air Temperature

The values of the air temperature (T_{air}) evolution obtained within the auditorium and the solar greenhouse are presented in Fig. 4. In Fig. 4, the evolution of the exterior T_{air} considered in the numerical simulations is also presented. The daily evolution

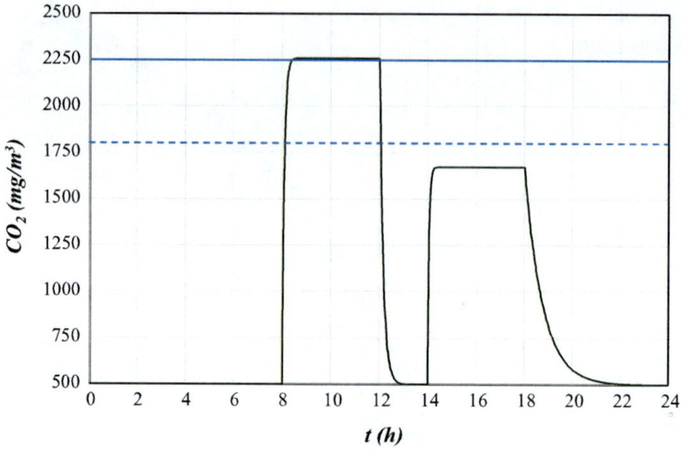

Fig. 3 CO_2 concentration evolution within the auditorium. The dashed and solid lines represent the acceptable limit proposed, respectively, in the international standard [17] and in the Portuguese standard [18]

of the exterior temperature represents a typical cold day of the winter season in the region of the location of the building and was obtained by a nearby weather station.

The T_{air} inside the solar greenhouse is always higher than the temperature outside, even at night, where the difference varies between +1.2 and +4.8 °C. During the day, this difference is more significant, reaching the maximum difference of 20.6 °C around 2 p.m. These results demonstrate that the use of heated air in the solar greenhouse and transferred to the interior of the auditorium adds to the increase in the T_{air} obtained there. The T_{air} inside the auditorium is, in general, higher than the

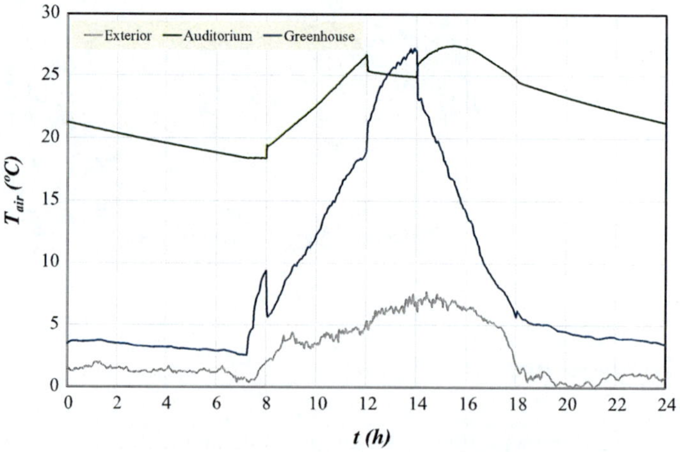

Fig. 4 Evolution of air temperature

T_{air} inside the solar greenhouse, except during the period of greatest sun exposure, between about 1 p.m. and 2 p.m. During the night, when there is no occupancy, the T_{air} inside the auditorium varies between 18.4 °C and 24.5 °C. During morning occupancy, the T_{air} inside the auditorium varies almost linearly between 19.4 and 26.7 °C. During the afternoon occupancy period, the T_{air} inside the auditorium oscillates between 24.5 and 27.5 °C (obtained around 3.30 p.m.). Thus, it can be seen that the solar greenhouse contributes positively to the heating of the air within the auditorium and to the maintenance of that T_{air} within a range of values suggested by the Portuguese standard [19], even at night.

3.3 TC Level

The TC level of occupants is estimated by the daily evolution of the PMV index whose results obtained throughout a day can be seen in Fig. 5. According to the results obtained, in 77% of the occupancy period it is verified that the occupants' thermal comfort level is acceptable by PMV index values at least within category C [13]: during the morning, mostly by negative values of the PMV index; during the afternoon (except the last 15 min), by positive values of the PMV index. As the PMV index also depends on the T_{air} indoors [12, 13], the positive contribution given by solar greenhouse to the rise in indoor T_{air} in the auditorium during the winter season is also positively observed in the improvement of TC-level conditions provided to the occupants.

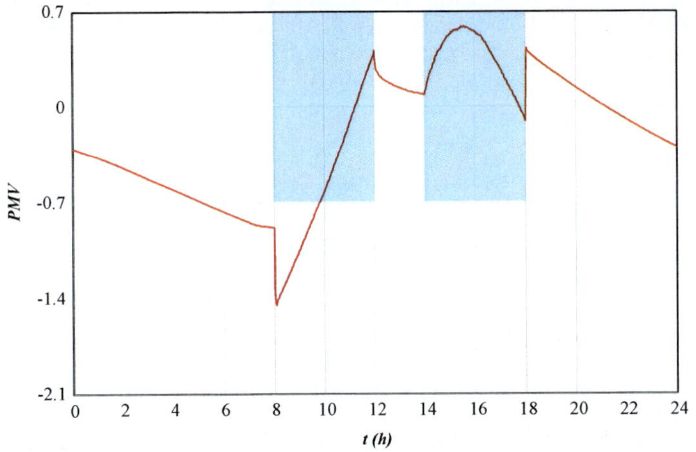

Fig. 5 Evolution of PMV index. The blue shaded areas represent, during occupancy, the acceptable limits of TC level according to category C of the standard [13]

4 Conclusions

In this article, we studied the contribution of a solar greenhouse, when attached to an auditorium, to improve the environmental conditions provided to the occupants of that auditorium. The study was performed numerically using the authors' own research software. The aforementioned environmental conditions evaluated were the IAQ level, through the CO_2 concentration verified inside the auditorium, and the TC level, through the PMV index. The numerical study was conducted for the winter season.

The use of solar greenhouse contributed positively to the increase in indoor air temperature in the auditorium. On the other hand, it was found that the ventilation process implemented ensured a suitable airflow rate for the number of occupants present in the auditorium. Therefore, the results obtained lead to the following main conclusions:

- It is possible to obtain, during the period of occupation, acceptable levels of IAQ, in the morning according to the Portuguese standard [18], in the afternoon according to the international standard [17];
- An acceptable TC level is provided to the occupants during 77% of the auditorium's occupancy time by values of the PMV index within category C of the ISO 7730 standard [13]. The time interval in which the TC level is unacceptable corresponds to the beginning of the morning when the heating process of the interior air of the auditorium by heated and ventilated air from the solar greenhouse begins.

Acknowledgements The authors would like to acknowledge the project (SAICT-ALG/39586/2018) from Algarve Regional Operational Program (CRESC Algarve 2020), under the PORTUGAL 2020 Partnership Agreement, through the European Regional Development Fund (ERDF) and the National Science and Technology Foundation (FCT).

References

1. Asa'd O, Ugursal V, Ben-Abdallah N (2019) Investigation of the energetic performance of an attached solar greenhouse through monitoring and simulation. Energy Sustain Dev 53:15–29
2. Aelenei D, Leal H, Aelenei L (2014) The use of attached-sunspaces in retrofitting design: the case of residential buildings in Portugal. Energy Procedia 48:1436–1441
3. Sánchez-Ostiz A, Monge-Barrio A, Domingo-Irigoyen S, González-Martínez P (2014) Design and experimental study of an industrialized sunspace with solar heat storage. Energy Build 80:231–246
4. Bataineh K, Fayez N (2011) Analysis of thermal performance of building attached sunspace. Energy Build 43(8):1863–1868
5. Owrak M, Aminy M, Jamal-Abad M, Dehghan M (2015) Experiments and simulations on the thermal performance of a sunspace attached to a room including heat-storing porous bed and water tanks. Build Environ 92:142–151

6. Conceição E, Nunes A, Gomes J, Lúcio M (2010) Application of a school building thermal response numerical model in the evolution of the adaptive thermal comfort level in Mediterranean environment. Int J Vent 9(3):287–304
7. Conceição E, Santiago C, Lúcio M, Awbi H (2018) Predicting the air quality, thermal comfort and draught risk for a virtual classroom with desk-type personalised ventilation systems. Buildings 8(2):35
8. Conceição E, Lúcio M (2010) Numerical study of the influence of opaque external trees with pyramidal shape on the thermal behaviour of a school building in summer conditions. Indoor Built Environ 19(6):657–667
9. Smith V, Sookoor T, Whitehouse K (2012) Modeling building thermal response to HVAC zoning. ACM SIGBED Rev 9(3):39–45
10. Guo J, Zheng W, Tian Z, Wang Y, Wang Y, Jiang Y (2022) The short-term demand response potential and thermal characteristics of a ventilated floor heating system in a nearly zero energy building. J Energy Storage 45:103643
11. Conceição E, Gomes J, Awbi H (2019) Influence of the airflow in a solar passive building on the indoor air quality and thermal comfort levels. Atmosphere 10(12):766
12. Fanger P (1970) Thermal comfort: analysis and applications in environmental engineering. Danish Technical Press, Copenhagen, Denmark
13. ISO 7730 (2005) Ergonomics of the thermal environments—analytical determination and interpretation of thermal comfort using calculation of the PMV and PPD indices and local thermal comfort criteria. International Standard Organization, Geneva, Switzerland
14. Conceição E, Silva M, Viegas D (1997) Air quality inside the passenger compartment of a bus. J Expo Anal Environ Epidemiol 7(4):521–534
15. Conceição E, Farinho J, Lúcio M (2012) Evaluation of indoor air quality in a classroom equipped with crossed ventilation. Int J Vent 11(1):53–67
16. Asif A, Zeeshan M, Jahanzaib M (2018) Indoor temperature, relative humidity and CO_2 levels assessment in academic buildings with different heating, ventilation and air-conditioning systems. Build Environ 133:83–90
17. ANSI/ASHRAE Standard 62-1 (2016) Ventilation for acceptable indoor air quality. American Society of Heating, Refrigerating and Air-Conditioning Engineers, Atlanta, GA, USA
18. Portaria n.º 353-A/2013 (2013) Regulamento de desempenho energético dos edifícios de comércio e serviços (RECS)—Requisitos de ventilação e qualidade do ar interior. Diário da República—1ª série 245, 6644-(2)-6644-(9)
19. Portaria n.º 353-D/2013 (2013) Regulamento de desempenho energético dos edifícios de comércio e serviços (RECS)—Requisitos de conceção para edifícios novos e intervenções. Diário da República—1ª série 233, 6628-(40)-6628-(72)

K-means Optimizer: An Efficient Optimization Algorithm for Predicting the Uncertain Material Parameters in Real Structures

Hoang-Le Minh, Thanh Sang-To, Magd Abdel Wahab, and Thanh Cuong-Le

Abstract In this paper, an efficient optimization algorithm is used to predict the uncertain material parameters based on finite element (FE) model updating. For this purpose, an objective function is determined based on the difference between the dynamic characteristics of the measurement and FE results. Then, this objective function is optimized to determine the model's uncertain material parameters. K-means optimizer (KO), a powerful algorithm developed by the authors in previous work, is used to handle this procedure. Then, the model's uncertain material parameters are determined by optimizing the objective function. The feature of KO is that it creates a strong balance between the abilities of exploration and exploitation. At the same time, this algorithm has also significantly improved the escape from local optima, which is appreciated to solve the complex problems. To evaluate the effectiveness of the proposed method, a practical example is used, i.e., a simple steel beam. The results obtained in this study have proved the effectiveness of the proposed method. At the same time, the KO algorithm is also shown to be a powerful algorithm for solving the optimization problems.

Keywords Metaheuristic optimization · K-means optimizer · FE model updating

1 Introduction

Accurate model development is a major focus of contemporary structural dynamics analysis. Many applications of civil engineering structures, including damage detection, structural health monitoring, structural control, structural evaluation, and assessment, use these precise models. Simplifying assumptions are frequently used while

H.-L. Minh (✉) · M. Abdel Wahab
Department of Electrical Energy, Metals, Mechanical Constructions, and Systems, Faculty of Engineering and Architecture, Ghent University, 9000 Gent, Belgium
e-mail: hoang.lm@ou.edu.vn; minhhoang.le@ugent.be

H.-L. Minh · T. Sang-To · T. Cuong-Le
Center for Engineering Application & Technology Solutions, Ho Chi Minh City Open University, Ho Chi Minh City, Vietnam

M. Abdel Wahab (ed.), *Proceedings of the 5th International Conference on Numerical Modelling in Engineering*, Lecture Notes in Civil Engineering 311,
https://doi.org/10.1007/978-981-19-8429-7_7

creating finite element (FE) models of structures. A structure's FE model is built using highly idealized engineering designs and specifications, which may or may not accurately depict all of the physical characteristics of a real structure. Field dynamic tests are used to confirm the analytical model, but it is obvious that their outcomes typically natural frequencies and mode shapes do not match those predicted by the model. These differences result from inaccurate boundary conditions, simplifying assumptions regarding structural geometry, or materials. The challenge of updating the analytical model based on dynamic measurements is known as FE model updating technique. Basically, the purpose of model updating is to modify the mass, stiffness, and uncertain material parameters of the numerical model in order to obtain better agreement between numerical and test results. Generally speaking, structural dynamics uses FE model update to adapt analytical models to experimental vibration measurements. It became clear in the 1990s that this was crucial to both mechanical and civil constructions. One can distinguish between direct and iterative techniques.

FE model updating based on the optimization algorithm is one of the effective techniques to solve the inverse problem. Accordingly, a highly reliable algorithm will be used to handle the objective functions. This technique has proven reliable and effective in solving highly complex FE models [1–5]. More and more optimization algorithms are being proposed as a result of the strong development in solving the optimization problems over the past two decades such as Teaching–Learning-Based Optimization (TLBO) [6], Wasp Swarm Algorithm (WSO) [7], Simulated annealing (SA) [8], Gravitational Local Search (GLSA) [9], efficient Planet Optimization Algorithm (POA) [10], etc. No Free Lunch Theorem of Optimization [11] proved that no optimization algorithm consistently outperformed others in a variety of optimization problems. It implies that metaheuristics work better for some optimization issues and less effectively for others. Therefore, choosing a suitable sub-algorithm for a particular problem becomes increasingly challenging especially for solving the FE model updating inverse problem.

2 K-means Optimizer

K-means optimizer (KO) is a recently developed new metaheuristic optimization algorithm [12]. This algorithm is to establish a new trend that is distinct from those of most other algorithms. The clustering problem and some important mathematical assumptions serve as the foundation of KO.

The balance in KO is secured by a parameter called "*threshold*". This is a turn parameter, and its value is defined by user according to a particular optimization problem. In KO, each search agent $\overrightarrow{SA_i^{(k)}} = \{sa_{i1}, sa_{i2}, \ldots, sa_{iD}\} \in \mathbb{R}^{1\times D}, (i = 1, 2, \ldots, N)$ is presented by a new term named "*Position density probability (PDP)*" as given in Eq. (1).

$$PDP_{SA_i} = CL_{SA_i} \sin\left(\frac{k\pi}{K_{\max}}\right) \tag{1}$$

where kth is the current iteration and $K_{\max}$ is the maximum number of iterations. $CL_{SA_i}(k)$ is called "*characteristic length*" as shown in Eq. (2).

$$\begin{aligned} D_{SA_i}(k) &= \frac{\left\| \overrightarrow{SA_i^{(k)}} - \overrightarrow{P_{best}^{(k)}} \right\|_2}{\left\| \overrightarrow{P_{worst}^{(k)}} - \overrightarrow{P_{best}^{(k)}} \right\|_2} \\ CL_{SA_i}(k) &= e^{-D_{SA_i}(k)} \end{aligned} \tag{2}$$

where $D_{SA_i}(k)$ is the relative distance of each search agent. It is calculated based on the positions of the best solution $\overrightarrow{P_{best}^{(k)}}$ and the worst solution $\overrightarrow{P_{worst}^{(k)}}$ recorded at each iteration.

Based on the value of the "*Position density probability (PDP)*" of each search agent $\overrightarrow{SA_i^{(k)}}$, its movement strategy in the next iteration is given in Algorithm 1.

Algorithm 1: The selection for movement strategy of each search agent

For *each search agent* $\overrightarrow{SA_i^{(k)}} = \{sa_{i1}, sa_{i2}, \ldots, sa_{iD}\} \in \mathbb{R}^{1\times D}, (i = 1, 2, \ldots, N)$

  IF $PDP_{SA_i} = CL_{SA_i} \sin\left(\frac{k\pi}{K_{\max}}\right) > threshold$;

  Move bias to the ability of exploitation

  Else

  Move bias to the ability of exploration

  End

End

2.1 The Movement Strategy for Exploitation

In KO, based on the K-means clustering algorithm, we can find the number of centroids defined by a potential matrix symbolized $\overline{P}$, which is in ascending order according to their objective function value, as given in Eq. (3).

$$\overline{P(k)} = \begin{bmatrix} \overline{p_{1,1}^{(k)}} & \overline{p_{1,2}^{(k)}} & \cdots\cdots & \overline{p_{1,D}^{(k)}} & \overline{Fitness_1^{(k)}} \\ \overline{p_{2,1}^{(k)}} & \overline{p_{2,1}^{(k)}} & \cdots\cdots & \overline{p_{2,D}^{(k)}} & \overline{Fitness_2^{(k)}} \\ \vdots & \vdots & \vdots\ \vdots & \vdots & \vdots \\ \vdots & \vdots & \vdots\ \vdots & \vdots & \vdots \\ \overline{p_{f(k),1}^{(k)}} & \overline{p_{f(k),2}^{(k)}} & \cdots\cdots & \overline{p_{f(k),D}^{(k)}} & \overline{Fitness_{f(k)}^{(k)}} \end{bmatrix}^{f(k)\times D} \tag{3}$$

where $f(k)$ is the Linear Population Size Reduction (LPSR) as given in the following Eq. (4):

$$f(k) = \frac{(N-4)k}{1-K_{\max}} + N - \frac{(N-4)}{1-K_{\max}} \tag{4}$$

where N is the number of initial population.

By using K-means algorithm, in which the number of centroids is selected as $K = 3$, we can find the centroid of clustering at each iteration. $\left\{\overrightarrow{C_{centroid}}\right\} = \left\{\overrightarrow{C_1^{(k)}}, \overrightarrow{C_2^{(k)}}, \overrightarrow{C_3^{(k)}}\right\}$. Especially the search space for establishing $\left\{\overrightarrow{C_{centroid}}\right\}$ will be shrunk as the number of iterations increases. Thus, at the last iterations, the position of $\left\{\overrightarrow{C_{centroid}}\right\}$ will be close to $\overrightarrow{P_{best}^{(k)}}$. To expand the potential search spaces, a new quantity position will be introduced, called mean centroid vector position $\overrightarrow{C_{mean}^{(k)}}$ as given in Eq. (5):

$$\overrightarrow{C_{mean}^{(k)}} = \frac{\overrightarrow{C_1^{(k)}} + \overrightarrow{C_2^{(k)}} + \overrightarrow{C_3^{(k)}}}{3} \tag{5}$$

Thus, at each iteration, we can record a potential search space $\Omega^* = \left\{\overrightarrow{C_1^{(k)}}, \overrightarrow{C_2^{(k)}}, \overrightarrow{C_3^{(k)}}, \overrightarrow{C_{mean}^{(k)}}, \overrightarrow{P_{best}^{(k)}}\right\}$. Based on this search spaces, three movement strategies are established to move the high-quality search space as follows:

For the first strategy: The search spaces between the current position $\overrightarrow{SA_i^{(k)}}$ and the best position $\overrightarrow{P_{best}^{(k)}}$, the mean centroid $\overrightarrow{C_{mean}^{(k)}}$ are considered as shown in Eq. (6).

$$\overrightarrow{SA_i^{(k+1)}} = w_{11}\overrightarrow{P_{best}^{(k)}} + d_1\left|\overrightarrow{SA_i^{(k)}} - w_{12}\overrightarrow{P_{best}^{(k)}}\right|$$

$$+ d_2\left|\overrightarrow{SA_i^{(k)}} - w_{13}\overrightarrow{C_{mean}^{(k)}}\right| \tag{6}$$

where w_{11} is a random scalar number, w_{12} and w_{13} are the random scalar vectors given in Eq. (7).

$$\begin{aligned} w_{11} &= 2rand(0, 1) \\ w_{12} &= rand(1, \text{dim}) \\ w_{13} &= rand(1, \text{dim}) \end{aligned} \tag{7}$$

where $d_1(k)$ and $d_2(k)$ are two scalar parameters called "*drift control*". Their values are reduced as the iterations increase as shown in Eq. (8).

$$d_1(k) = (r_1 f_{d1}(k) + r_2 f_{d2}(k))$$

$$d_2(k) = (r_3 f_{d1}(k) + r_4 f_{d2}(k)) \tag{8}$$

where $f_{d_1}(k)$ and $f_{d_2}(k)$ are two symmetric functions having values bounded in range $[-2,\ 2]$ as shown in Eq. (9).

$$\begin{aligned} f_{d1}(k) &= 2\sin\left[0.1K_{\max}\left(1 - \frac{k}{K_{\max}}\right)\right]\left(1 - \frac{k}{K_{\max}}\right) \\ f_{d2}(k) &= -2\sin\left[0.1K_{\max}\left(1 - \frac{k}{K_{\max}}\right)\right]\left(1 - \frac{k}{K_{\max}}\right) \end{aligned} \tag{9}$$

For the second strategy: This strategy finds the region close to the high-quality search space $\Omega^* = \left\{\overrightarrow{C_1^{(k)}},\ \overrightarrow{C_2^{(k)}},\ \overrightarrow{C_3^{(k)}},\ \overrightarrow{C_{mean}^{(k)}},\ \overrightarrow{P_{best}^{(k)}}\right\}$ as given in Eq. (10).

$$\begin{aligned} \overrightarrow{SA_i^{(k+1)}} =& w_{21}\overrightarrow{SA_i^{(k)}} + w_{22}\left(\overrightarrow{P_{best}^{(k)}} - w_{23}\overrightarrow{SA_i^{(k)}}\right) \\ &+ \sum_{k=1}^{3}\delta_k\left(\overrightarrow{C_k^{(k)}} - \beta_k\overrightarrow{SA_i^{(k)}}\right) \end{aligned} \tag{10}$$

where w_{21} is the same values with w_{11} in case of the first trend. w_{22}, w_{23}, and β_k are given in Eq. (11).

$$\begin{aligned} w_{22} &= rand(0, 1) \\ w_{23} &= rand(0, 1) \\ \beta_{k,k=1,2,3} &= rand(0, 1) \end{aligned} \tag{11}$$

For the third strategy: The search space between the mean centroid vector $\overrightarrow{C_{mean}^{(k)}}$ and the best position is the destination of each search agent $\overrightarrow{SA_i^{(k)}}$ as Eq. (12).

$$\overrightarrow{SA_i^{(k+1)}} = w_{31}\overrightarrow{C_{mean}^{(k)}} + w_{32}\left(\overrightarrow{P_{best}^{(k)}} - w_{33}\overrightarrow{C_{mean}^{(k)}}\right) \tag{12}$$

where $w_{31},\ w_{32},\ w_{33}$ are the same as $w_{11},\ w_{12},\ w_{13}$ in the first trend.

2.2 The Movement Strategy for Exploration

The ability of exploration is established based on two movement strategies.

For the first strategy: This strategy uses the position of mean centroid vector $\overrightarrow{C_{mean}^{(k)}}$ and a modification step length S to create a wide range search space as given in Eq. (13).

$$\overrightarrow{SA_i^{(k+1)}} = \overrightarrow{C_{mean}^{(k)}} + \text{step length}(S) \quad IF\ rand(0,1) > 0.5 \tag{13}$$

where step length "S" is obeyed by Lévy distribution as given in Eq. (14)

$$\begin{aligned} S &= \frac{U}{|V|^{1/\beta}} \\ \beta &= 1 + rand(0,1) \\ U &= \text{normal}\left(0, \sigma_u^2\right) \\ V &= \text{normal}\left(0, \sigma_v^2\right) \\ \sigma_u &= \left\{\frac{\Gamma(1+\beta)\sin(\pi\beta/2)}{\Gamma[(1+\beta)/2]\beta 2^{(\beta-1)/2}}\right\}^{1/\beta}, \sigma_v = 1,\ 1 \le \beta \le 2 \end{aligned} \tag{14}$$

For the second strategy: If the global optimal locates in the search space, which is far from $\overrightarrow{C_{mean}^{(t)}}$ and if Eq. (13) cannot reach the better solution. The new search space will be established to replace the old ones as given in Eq. (15).

$$\overrightarrow{SA_i^{(k+1)}} = L_b + rand(0,1)(U_b - L_b) \quad IF\ rand(0,1) < 0.5 \tag{15}$$

where L_b and U_b are the lower-bound and upper-bound vectors, respectively.

3 Numerical Examples

3.1 The Objective Function

To establish the objective function, the error between the frequencies obtained from FE model analysis and the frequencies obtained from measurement is considered. Thus, the objective function, which is called "*Root mean square error*", is given in Eq. (16).

$$R = \frac{1}{n_\omega}\sqrt{\sum_{i=1}^{n_\omega}(f(d) - f(\alpha))^2} \tag{16}$$

where $f(d)$ is the frequency at damaged stage obtained by test real structure, $f(\alpha)$ is the frequency obtained from FE model updating. n_ω is the number of frequencies, which is used to calculate the objective function. The procedure of minimizing the objective function will be handled by KO. The best solution, which makes an acceptable error, can be accepted.

3.2 The Steel Beam Test

The first example is implemented based on the test of Samir et al. [13]. This test has details of geometry and set-up given in Fig. 1, and the Finite element (FE) model is illustrated in Fig. 2.

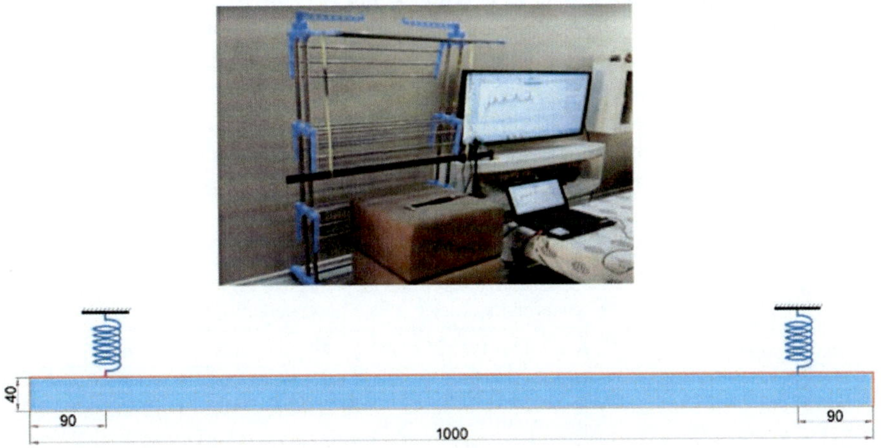

Fig. 1 Experiment setup and the geometry of experimental steel beam

Fig. 2 The FE model of steel beam test

Table 1 The initial material parameters and the frequencies measurements

The assumption of initial material parameters	Value
Mass density (kg/m^3)	7850
Elasticity (GPa)	210
Stiffness of Spring 1	Unknown
Stiffness of Spring 2	Unknown
Frequencies measurements	
Mode 1	211.33
Mode 2	578.52
Mode 3	1121.7
Mode 4	1821.1

The assumption of initial material parameters and the values of frequencies obtained by measurement are shown in Table 1.

Mass density, modulus of elasticity, and stiffness of the two springs are the variants, and it can be converted to a solution vector $\overrightarrow{X} = (x_1, x_2, x_3, x_4)$. The inverse problem handled by KO is used to find the best solution of $\overrightarrow{X} = (x_1, x_2, x_3, x_4)$ based on the process of FE model updating. For this purpose, the set-up of KO including $N = 10$ (the population size) and $K_{\max} = 150$ (maximum iterations) is generated in the initial stage.

In order to limit the search space of uncertain material parameters, the boundary condition of the problem is set up as given in Table 2.

The process of finding the best solution using KO optimization algorithm is illustrated in Fig. 3. And the results obtained are shown in Tables 3 and 4. The historical of searching process of the objective function, uncertain material parameters, and the frequencies obtained over the course of iterations are shown in Fig. 4. The results demonstrated that KO optimizer can find the suitable solution with acceptable error over the course of only 150 iterations. The largest error is recorded in the third mode with a value of 3.66%.

Table 2 The boundary condition of uncertain material parameters

Material parameter	Lower condition	Upper condition
Mass density (kg/m3)	7600	8000
Elasticity (GPa)	180	220
Spring stiffness (kN/m)	10^3	10^4

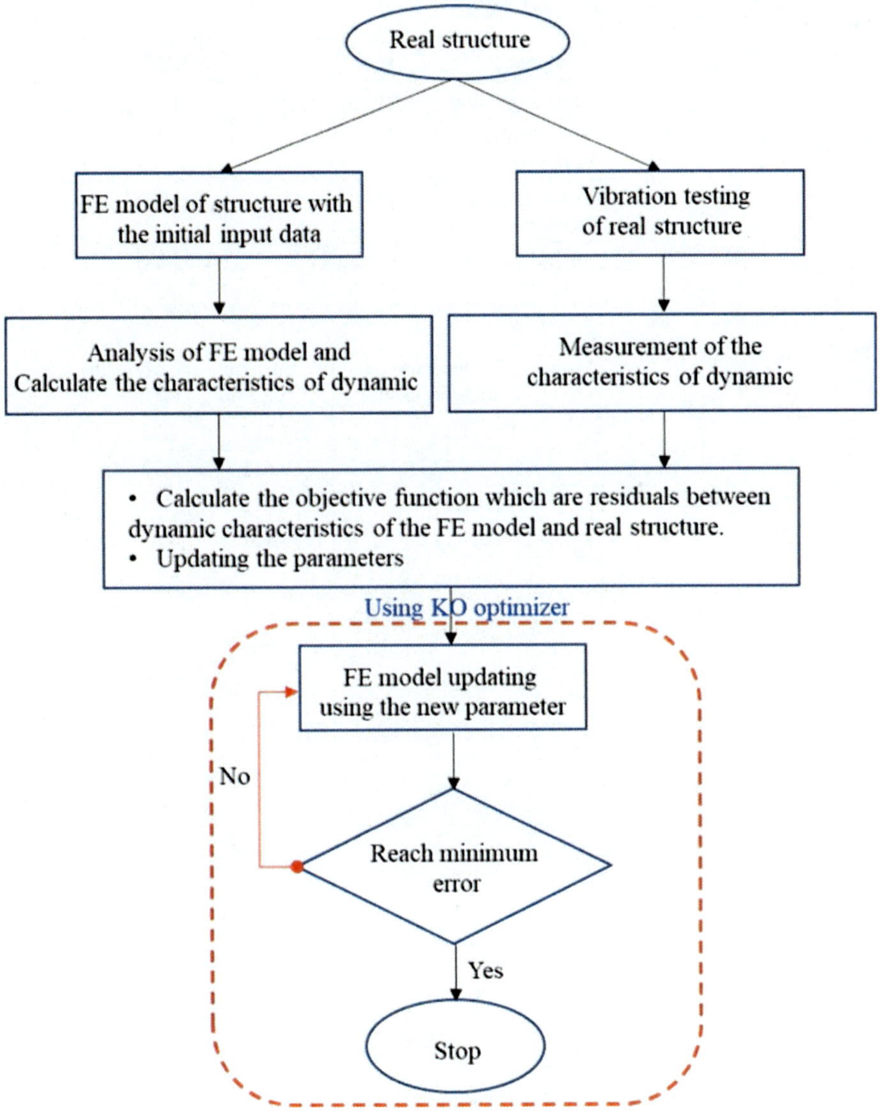

Fig. 3 The process of finding the best solution using FE model updating handled by KO optimizer

4 Conclusion

This paper introduces K-means optimizer (KO), an efficient optimization algorithm for predicting uncertain material parameters in simple steel structures. The FE model updating method with experimental modal data is used in two steps. To obtain a reference model, the initial FE model is tuned to the undamaged state in the first

Table 3 The best solution exploited using KO

Uncertain material parameters	The initial values	The values exploited by KO
Elasticity (GPa)	210	182.034
Mass density (kg/m^3)	7850	8549.254
Stiffness of spring 1 (kN/m)	Unknown	6099.025
Stiffness of spring 2 (kN/m)	Unknown	6262.583

Table 4 The frequencies obtained using KO

Mode	Frequencies (measurements)	Frequencies (Using KO)	Error using KO (%)
1	211.33	211.233	0.25
2	578.52	569.611	0.92
3	1121.7	1156.815	3.66
4	1821.1	1798.192	0.89

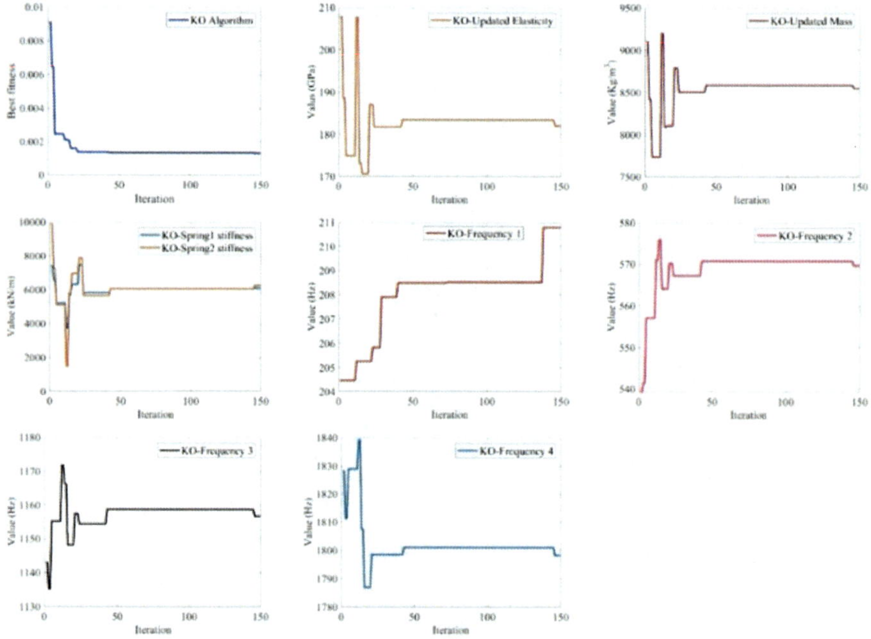

Fig. 4 The results obtained by KO over the 150 iterations

step. In the second step, the reference model is updated based on the new parameters generated by KO's movement strategies. This process is repeated indefinitely until the minimum value of the objective function is found. The method is illustrated using simple steel beam tests, where the applied KO successfully identifies the uncertain material parameters with acceptance error.

Acknowledgements The authors acknowledge the financial support of VLIR-UOS TEAM Project, VN2018TEA479A103, "Damage assessment tools for Structural Health Monitoring of Vietnamese infrastructures," funded by the Flemish Government.

vliruos SHARING MINDS, CHANGING LIVES With the support of THE BELGIAN DEVELOPMENT COOPERATION .be

References

1. Minh H-L, Khatir S, Abdel Wahab M, Cuong-Le T (2021) An enhancing particle swarm optimization algorithm (EHVPSO) for damage identification in 3D transmission tower. Eng Struct 242:112412. https://doi.org/10.1016/j.engstruct.2021.112412
2. Minh H-L, Khatir S, Rao RV, Abdel Wahab M, Cuong-Le T (2021) A variable velocity strategy particle swarm optimization algorithm (VVS-PSO) for damage assessment in structures. Eng Comput. https://doi.org/10.1007/s00366-021-01451-2
3. To TS, Le MH, Danh T-T, Khatir S, Abdel Wahab M, Le TC (2022) Combination of intermittent search strategy and an improve particle swarm optimization algorithm (IPSO) for model updating. Frattura ed Integrita Strutturale—Fract Struct Integrity 16(59):141–152
4. Sang-To T, Hoang-Le M, Abdel Wahab M, Cuong-Le T (2022) Predicting the displacement of diaphragm wall for deep excavation problem on the basing thickly soft soil in an urban area using semi-top-down construction method. In: Proceedings of the 2nd international conference on structural damage modelling and assessment. Springer, pp 49–56
5. Cuong-Le T, Minh H-L, Khatir S, Wahab MA, Tran MT, Mirjalili S (2021) A novel version of Cuckoo search algorithm for solving optimization problems. Expert Syst Appl 186:115669
6. Rao RV, Savsani VJ, Vakharia D (2011) Teaching–learning-based optimization: a novel method for constrained mechanical design optimization problems. Comput Aided Des 43(3):303–315
7. Pinto PC, Runkler TA, Sousa JM (2007) Wasp swarm algorithm for dynamic MAX-SAT problems. International conference on adaptive and natural computing algorithms. Springer, pp 350–357
8. Kirkpatrick S, Gelatt CD, Vecchi MP (1983) Optimization by simulated annealing. Science 220(4598):671–680
9. Webster B, Bernhard PJ (2003) A local search optimization algorithm based on natural principles of gravitation
10. Sang-To T, Hoang-Le M, Wahab MA, Cuong-Le T (2022) An efficient Planet Optimization Algorithm for solving engineering problems. Sci Rep 12(1):1–18
11. Wolpert DH, Macready WG (1997) No free lunch theorems for optimization. IEEE Trans Evol Comput 1(1):67–82. https://doi.org/10.1109/4235.585893
12. Minh H-L, Sang-To T, Wahab MA, Cuong-Le T (2022) A new metaheuristic optimization based on K-means clustering algorithm and its application for structural damage identification in a complex 3D concrete structure. Knowl-Based Syst 109189
13. Khatir S et al (2020) An efficient hybrid TLBO-PSO-ANN for fast damage identification in steel beam structures using IGA. Smart Struct Syst 25(5):605–617

A Nonlinear Approach to Investigate the Effect of Sheet Pile Toe's Embedded Length on the Lateral Displacement Derived from Soft Clay-Deep Excavation

Thanh Sang-To, Minh Hoang-Le, Quoc Thien Huynh, Magd Abdel Wahab, and Thanh Cuong-Le

Abstract In this paper, a nonlinear approach is employed to investigate how the embedded length of sheet pile toe in soft clay affects the lateral displacement of retaining wall for a soft clay-deep excavation. Firstly, an inverse analysis using a finite element (FE) model and the Geotechnical software (PLAXIS), combined with field measurements is adopted to validate soil parameters. Then, a series of investigations are conducted to assess the effect of the sheet pile toe's embedded length on the displacement of retaining wall in detail. Finally, several conclusions can be drawn based on the obtained results.

Keywords Soft soil-deep excavation · Sheet piles toe's embedded length · Lateral displacement

1 Introduction

With the development of Science and Technology, together with the growing demand of humans, the underground space has been exploited to be used as museums, restaurants, parking lots, houses, etc. For this reason, a series of methods are employed to deal with deep excavation problems. One of them is steel sheet piling taking on the role of retaining wall [1]. Deep excavation using sheet piles has several advantages over general method, and one of the many advantages of this technique is environmental friendliness. Steel sheet piling can be reused many times for many different projects instead of single-use approach. That leads to more savings and

T. Sang-To (✉) · M. Abdel Wahab
Soete Laboratory, Faculty of Engineering and Architecture, Ghent University, Technologiepark Zwijnaarde 903, B-9052 Zwijnaarde, Belgium
e-mail: sang.tt@ou.edu.vn

T. Sang-To · M. Hoang-Le · T. Cuong-Le
Center for Engineering Application & Technology Solutions, Ho Chi Minh City Open University, Ho Chi Minh City, Vietnam

Q. T. Huynh
Institute of Research and Development, Duy Tan University, Danang 550000, Vietnam

M. Abdel Wahab (ed.), *Proceedings of the 5th International Conference on Numerical Modelling in Engineering*, Lecture Notes in Civil Engineering 311,
https://doi.org/10.1007/978-981-19-8429-7_8

less pollution for the construction of projects. Furthermore, sheet piles method also shows high strength when constructing is simple. At the same time, it needs very few workers to operate the sheet piles installation. A small team, only two to three workers, can do this work well on the construction site. In addition, steel sheet piling is able to apply for many distinct problems in practice, namely dam, embankment, slope stabilization, constructing bridges, etc.

To overcome Geotechnical engineering problems, various solutions were adopted from simple approach (numerical methods) to complex method (e.g. Artificial Intelligence). However, FE model plays an essential role in providing the effective solutions for real problems. The FE method using Geotechnical software, such as PLAXIS [2–4] or ABAQUS [5–7], etc., is a powerful tool to assess and predict the risks that can happen in Geotechnical engineering, e.g. in deep excavation problems [8, 9].

In this work, we present a nonlinear analysis to assess soil parameters using sheet piles under construction. Then, various investigations are conducted to determine the link between the toe's embedded length and retaining wall displacement. This study is divided into five sections. The next section will illustrate a real building for simulation. Section 3 presents an inverse analysis to determine soil parameters. While the key results are shown in the Sect. 4, conclusions are drawn in Sect. 5.

2 The Golden Star

To address the problem, the Golden Star (GOS) is used to assess this approach. With two basements, 4500 m^2, and 26 stories, this project is a kind of popular building in Ho Chi Minh city. In addition, GOS lies at the heart of the biggest city in Vietnam as shown in Fig. 1, and is constructed on a very soft clay layer having 20 m thick. The bottom-up method is used to excavate construction for GOS basements.

With 18 m length, steel sheet piling is selected is SP-IV (see Fig. 2) as a retaining wall of the excavation process. The construction process is listed in Table 1. There are three main excavation stages at depth −1.3, −4.05 and −6.65 m. Under excavation, two steel bracing systems, at −0.8 and 3.55 m are installed to support the retaining wall.

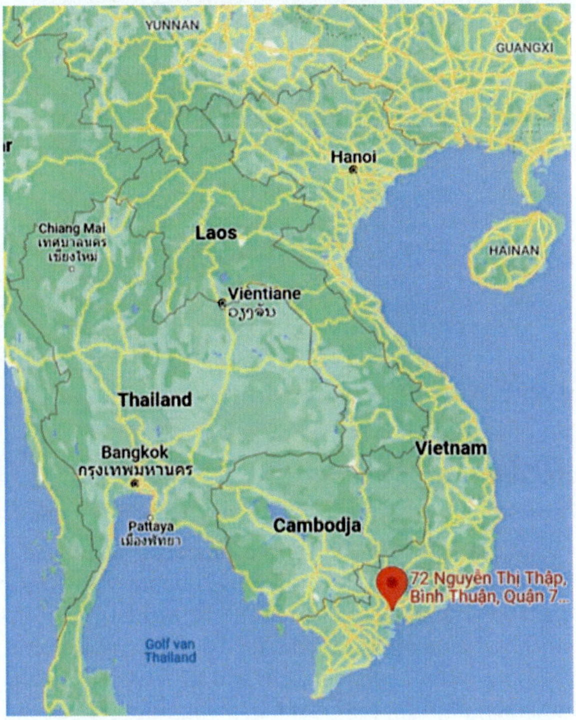

Fig. 1 Location of the GOS

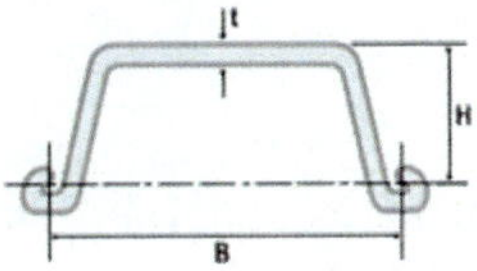

14. Sheet Piles 강널말뚝

Dimensions and Sectional Properties 치수 및 단면성능 KS F 4604

Section Type	표준 단면 치수 Dimension			단면적 Sectional Area		단위무게 Unit Weight		단면 2차 모멘트 Moment of Inertia		단면계수 Modulus of Section	
	B	H	t	Per Pile	Per Wall Width	Per Pile	Per Wall Width	Per Pile	Per Wall Width	Per Pile	Per Wall Width
	mm	mm	mm	cm^2	cm^2/m	kg/m	kg/m^2	cm^4	cm^4/m	cm^3	cm^3/m
SP-II	400	100	10.5	61.2	153.0	48.0	120	1,240	8,740	152	874
SP-III	400	125	13.0	76.4	191.0	60.0	150	2,220	16,800	223	1,340
SP-III A(E)	400	150	13.1	74.4	186.0	58.4	146	2,790	22,800	250	1,520
SP-III A(D)	400	150	13.0	76.4	191.0	60.0	150	3,060	22,600	278	1,510
SP-IV	400	170	15.5	96.9	242.5	76.1	190	4,670	38,600	362	2,270
SP-V M	500	175	16.5	109.6	219.2	86.0	172	5,345	40,290	383	2,310
SP-V A	500	200	19.5	133.8	267.6	105.0	210	7,960	63,000	520	3,150
SP-II w	600	130	10.3	78.7	131.2	61.8	103	2,110	13,000	203	1,000
SP-III w	600	180	13.4	103.9	173.2	81.6	136	5,220	32,400	376	1,800

Fig. 2 Parameters of steel sheet pile

Table 1 Main excavation steps of GOS

Construction sequence	Main steps
Stage 1	Install steel sheet piling
Stage 2	Excavate soil to −1.3 m
Stage 3	Install the first steel bracing system H350
Stage 4	Excavate soil to −4.05 m
Stage 5	Install the second steel bracing system H400
Stage 6	Excavate soil to −6.65 m

3 Back Analyses of Deep Excavation

3.1 Soil Model and Parameters

To deal with this problem, one of the advanced models is employed, i.e. hardening soil model (HSM). The hardening soil model is used popularly to forecast the lateral displacement of the soil retaining structures. The behaviour of soils is modelled in detail, in which, stiffness parameters are described based on power law. Basic feature of the present HSM is the stress dependency of soil stiffness. For conditions of strain and stress, the simulation implies, for example, the relationship as shown in Eq. (1)

$$E = E^{ref}\left(\frac{\sigma}{p^{ref}}\right)^{m} \quad (1)$$

where p^{ref}, default normally equal to 100 units, is a reference stress.

In this respect, Hardening soil model generally consists of three main soil stiffness parameters, E_{50}, E_{oed}, and E_{ur}, respectively, plastic straining due to primary deviatoric loading, and primary compression and the elastic un/reloading modulus. In this soil model, some parameters, such as cohesion c, friction angle φ and dilatancy angle ψ, together with three principal stresses, describes the limiting states of stress by a yield surface, and it is a hexagon as it can be seen in Fig. 3.

GOS is built on a very soft soil, 20 m thick, with the value of N_{spt} equal zero. The soft clay is a group, including three layers: 1a, 1b and 1c. All of them share the same soil characterization, nevertheless, there is a different modulus stiffness between them. Underneath the soft soil, a clay layer is stiff, 10 m thick, and the final layer is sand. Table 2 shows the main soil parameters of stratigraphy GOS. Some stiffness parameters such as E_{ur}^{ref} and E_{oed}^{ref} are chosen following the default setting used in PLAXIS. A model is built by PLAXIS 2D version 2020, simulating the construction process in detail as shown in Table 1. Figure 4 presents this process, including installation of retaining wall using steel sheet piling, removing soil, installation of the steel bracing systems.

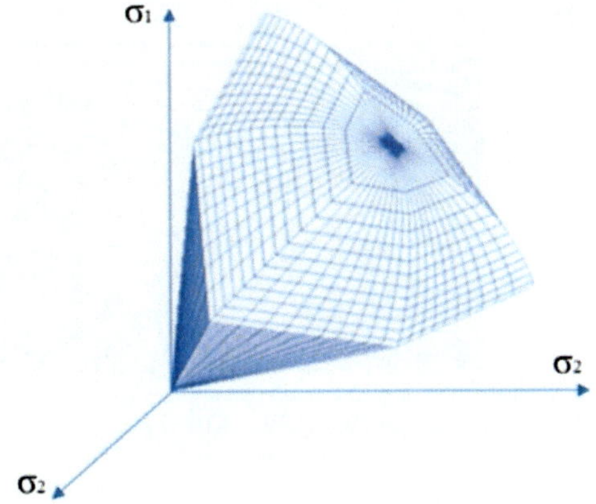

Fig. 3 The HSM yield surface in principal stress space with zero cohesion

Table 2 Soil data of GOS

Layer name	Soil layer	φ (°)	c (kPa)	E_{50}^{ref} (kPa)	m
0	Sand fill	25	5	15,000	0.5
1a	Soft soil	0	23.31	2000	0.9
1b	Soft soil	0	20.61	5000	0.9
1c	Soft soil	0	30.5	10,600	0.9
2	Stiff clay	16.6	28.2	60,000	0.75
3	Hard sand	27.2	16.7	40,000	0.5

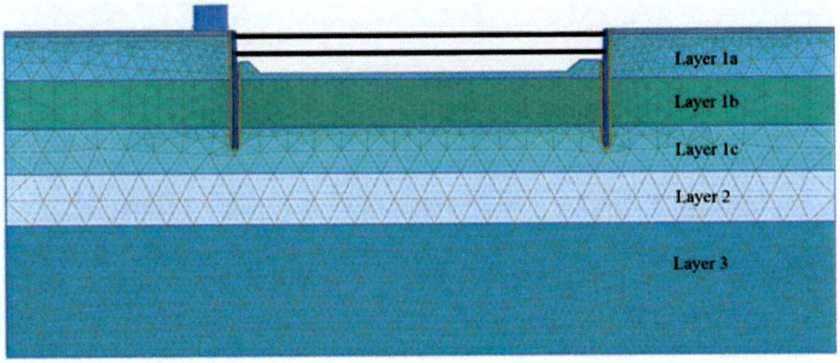

Fig. 4 Simulation under construction

3.2 *Results of Back Analyses*

The obtained results from back analyses are shown in Fig. 5. We can see that the maximum lateral displacement is 180 mm at the level of −7.0 m from the surface ground. Figure 6 illustrates a comparison between the achieved results using the

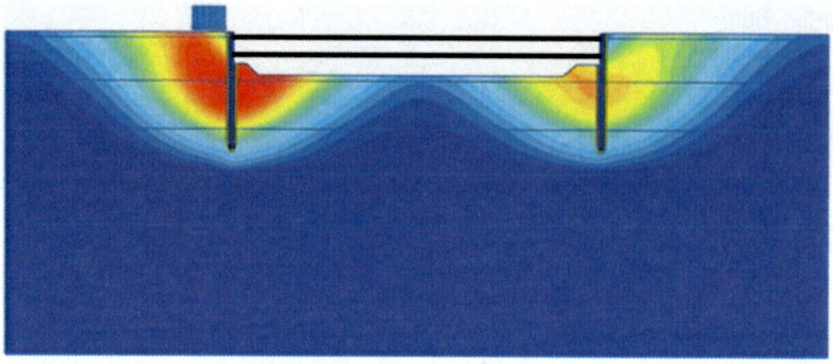

Fig. 5 Displacement using PLAXIS

FE model and field observations. It is interesting to note that information gathered from inverse analyses provided a proper estimation of displacement versus on-site measurement. Indeed, the difference between the two approaches is 180 mm (measurement) versus 189 mm (FE model), which is only 5%. These results are enough to prove that back analyses are reliable. For this reason, we will employ the soil characterizations to investigate the effect of sheet pile length on the lateral displacement of retaining wall.

4 Effect of Sheet Pile Toe's Embedded Length on the Displacement of Retaining Wall

To investigate the link between the effect sheet pile toe's embedded length and lateral displacement of retaining wall, a nonlinear approach using four scenarios with the different lengths of sheet piling is considered. Table 3 presents steel sheet piling length for each scenario.

Generally speaking, the results reveal that there is a link between displacement of retaining wall and length of sheet piles. Figure 7 presents lateral displacement of steel sheet piling for each scenario in detail. In the first case, the value of displacement is the largest. Specially, this position is located at the toe of retaining wall. This means that the sheet piles are too short. In other words, it is short compared with the requested one from depth excavation. Meanwhile, with a length of 12 m in the second case, the obtained result is improved versus the first case, however, the lateral displacement of retaining wall is also located at the toe of sheet piles. In the third scenario, it is the other way around, with 15 m in length the maximum displacement is currently located in the mid of the retaining wall. Obviously, based on the results presented herein that the length of sheet piles has a significant impact on lateral displacements of retaining wall.

Nevertheless, these results also cannot prove a linear relationship between the sheet pile toe's embedded length and displacement of retaining wall. This is illustrated in the rest of the cases in detail. In comparison with length of 18 m, which was proven

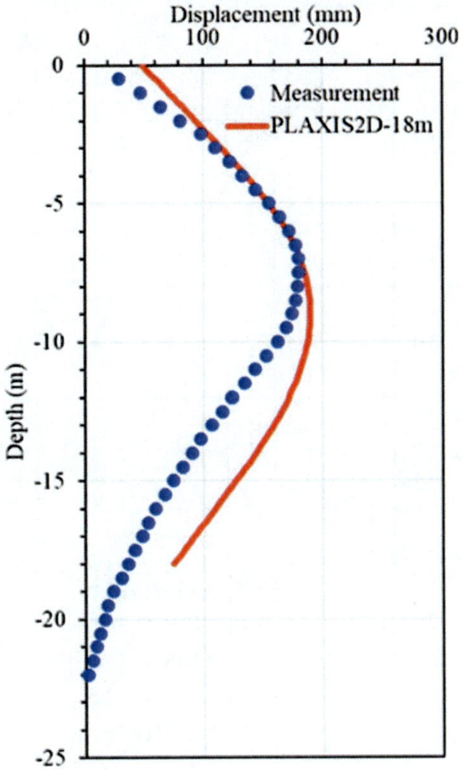

Fig. 6 Lateral displacement of retaining wall

Table 3 Length of steel sheet piling

Scenario	Total length—L (m)	Toe's embedded length—D (m)
1	9	2.35
2	12	5.35
3	15	8.35
4	21	14.35

that the results obtained properly from back analyses using on-site measurement and FE model, the final scenario shows that the difference between length of 18 and 24 m is insignificant for the lateral displacement of retaining wall. In this respect, once the length (L) gets 18 m, K ratio ($K = D/H$, see Fig. 8) is equal to 1.71. While $L = 21$, then the value of K reaches 2.16. Based on these results, we assume that length of sheet piles should reach a value of at least $K = 2$ to optimize the lateral displacement.

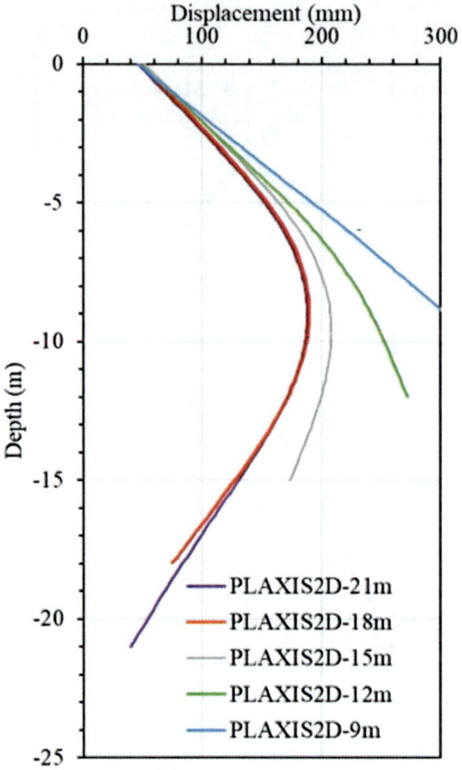

Fig. 7 Lateral displacement of retaining wall for scenarios

Fig. 8 Length of sheet piles into soil

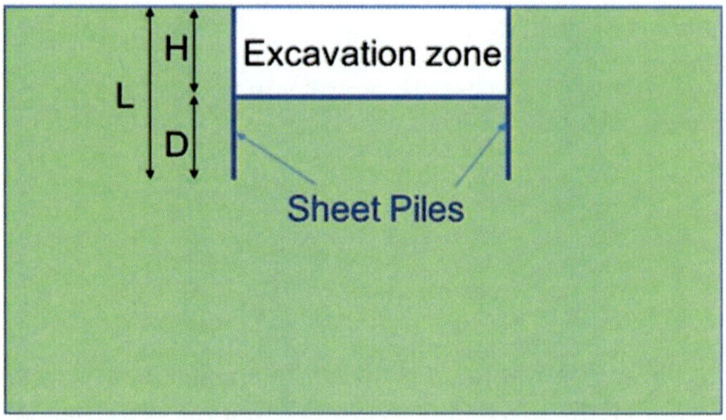

5 Conclusion

This paper presented a nonlinear simulation using HSM. A set of soil parameters is assessed via inverse analyses. The analyses using PLAXIS show good agreement between finite element analysis and on-site measurements. The authors conducted tests to determine the effect of K ratio on the displacements of retaining wall. By doing that, we found that $K = 2$ is a proper value for deep excavation problems using steel sheet piling rested on very soft clay.

Acknowledgements The authors acknowledge the financial support of VLIR-UOS TEAM Project, VN2018TEA479A103, 'Damage assessment tools for Structural Health Monitoring of Vietnamese infrastructures' funded by the Flemish Government.

References

1. Huynh QT, Lai VQ, Boonyatee T, Keawsawasvong S (2022) Verification of soil parameters of hardening soil model with small-strain stiffness for deep excavations in medium dense sand in Ho Chi Minh City, Vietnam. Innov Infrastruct Solut 7(1):1–20
2. Sang-To T, Hoang-Le M, Khatir S, Mirjalili S, Wahab MA, Cuong-Le T (2021) Forecasting of excavation problems for high-rise building in Vietnam using planet optimization algorithm. Sci Rep 11(1):1–10
3. Sang-To T, Hoang-Le M, Abdel Wahab M, Cuong-Le T (2022) Predicting the displacement of diaphragm wall for deep excavation problem on the basing thickly soft soil in an urban area using semi-top-down construction method. In: Proceedings of the 2nd international conference on structural damage modelling and assessment. Springer, pp 49–56
4. To TS, Le MH, Tran MV, Wahab MA, Thanh C-L (2022) Estimation displacement of diaphragm wall using hardening soil versus Mohr-Coulomb model. In: Recent advances in structural health monitoring and engineering structures. Springer, pp 345–351
5. Helwany S (2007) Applied soil mechanics with ABAQUS applications. Wiley
6. Cuong-Le T, Le-Minh H, Sang-To T (2022) A nonlinear concrete damaged plasticity model for simulation reinforced concrete structures using ABAQUS. Frat Ed Integrità Strutt 59:232–242
7. Minh H-L, Khatir S, Wahab MA, Cuong-Le T (2021) A concrete damage plasticity model for predicting the effects of compressive high-strength concrete under static and dynamic loads. J Build Eng 44:103239
8. Huynh QT, Lai VQ, Boonyatee T, Keawsawasvong S (2021) Behavior of a deep excavation and damages on adjacent buildings: a case study in Vietnam. Transp Infrastruct Geotechnol 8(3):361–389
9. Huynh QT, Lai VQ, Tran VT, Nguyen MT (2020) Analyzing the settlement of adjacent buildings with shallow foundation based on the horizontal displacement of retaining wall. Geotechnics for sustainable infrastructure development. Springer, pp 313–320

Damage Detection in a 3D Truss Structure Using Natural Frequencies and Metaheuristic Algorithms

Thanh Sang-To, Minh Hoang-Le, Magd Abdel Wahab, and Thanh Cuong-Le

Abstract In this study, a highly efficient optimization methodology for damage identification in steel structures, especially truss structures, is introduced. Natural frequencies are adopted as an objective function for optimization. In order to deal with the optimization problem, a set of optimization algorithms is employed for the purpose of damage detection in a three-dimensional (3D) truss structure. To assess the behaviour of truss structures, a finite element (FE) model is used. The efficiency of each algorithm is assessed through examples of truss-like structures. The achieved results are compared with each other, and based on these results, several conclusions are drawn.

Keywords Optimization · Damage identification · Damage detection · Truss

1 Introduction

During its cycle life, a structure undergoes many impacts of the environment, and the harmful effects of nature, such as storm, temperature, wind, earthquake, etc. Therefore, the assessment of the current structural health of a structure brings great significance not only for the purpose of life safety, but also for the environment. Structural Health Monitoring (SHM) provides data on its state and information. Thereby, the structure will be monitored for regular maintenance or reparation, even everyday if possible. It is essential to keep the structure in a safe state.

A truss is a popular structure, which is present in almost all fields of life. Therefore, developing a method for assessing SHM of truss structures is important for both practical applications and academic studies. It's a fact that prediction of SHM is

T. Sang-To (✉) · M. Abdel Wahab
Soete Laboratory, Faculty of Engineering and Architecture, Ghent University, Technologiepark Zwijnaarde 903, 9052 Zwijnaarde, Belgium
e-mail: sang.tt@ou.edu.vn

T. Sang-To · M. Hoang-Le · T. Cuong-Le
Center for Engineering Application & Technology Solutions, Ho Chi Minh City Open University, Ho Chi Minh City, Vietnam

M. Abdel Wahab (ed.), *Proceedings of the 5th International Conference on Numerical Modelling in Engineering*, Lecture Notes in Civil Engineering 311,
https://doi.org/10.1007/978-981-19-8429-7_9

usually based on a combination of on-site measurement and simulation using FE analysis. In recent years, many scholars have published many studies related to this field [1–3], in which methods using intelligent algorithms bring many positive results.

With the dramatic development of metaheuristic algorithm [4] in the last two decades, a series of algorithms have been proposed [2, 5–7], namely Grey Wolf Optimizer (GWO) [8], K-means clustering algorithm (KO) [9], Arithmetic Optimization Algorithm (AOA) [10], to deal with a lot of optimization problems.

In this study, Planet Optimization Algorithm (POA) [11] is used to predict the SHM of truss structure. The rest of this paper is structured as follows. Section 2 introduces the planet optimization algorithm. Section 3 describes the 3D truss system and evaluates the SHM, and final Sect. 4 draws some conclusions from the results presented in this paper.

2 Planet Optimization Algorithm

The Planet optimization algorithm is a metaheuristic algorithm inspired by the universal gravitational laws of Isaac Newton. POA is designed based on modelling the motion of objects in the universe. The fundamental principles of POA [11] are presented briefly as follows:

$$\left|\vec{F}\right| = G\frac{m_1 m_2}{R^2} \tag{1}$$

where F indicates the gravitational force; G is constant and illustrates the gravity factor. Meanwhile, R describes the distance between two planets and m presents the mass of the planet.

$$\left|\vec{M}\right| = \left|\vec{F}\right| R_{ij} = G\frac{m_i m_j}{R_{ij}^2} R_{ij} \tag{2}$$

- $m_i, m_j = \frac{1}{a^{obj_{/\alpha}}}$.

where $a = 2$ is a constant parameter, and $\alpha = \left|\max(obj) - obj_{Sun}\right|$. And $obj_{i,j}$, $\max(obj)$, obj_{Sun} are, respectively, the value of objective function of the ith or jth planet, the value of worst planet and the value of the Sun.

$$R_{ij} = \left\| X_i^t - X_j^t \right\| = \sqrt{\sum_{k=1}^{Dim} \left(X_i^t - X_j^t \right)^2} \tag{3}$$

For Global search

$$\overrightarrow{X_i^{t+1}} = \overrightarrow{X_i^t} + b \times \beta \times r_1 \times \left(\overrightarrow{X_{Sun}^t} - \overrightarrow{X_i^t}\right) \tag{4}$$

where $\overrightarrow{X_i^t}$ illustrates the current location of a planet ith in the iteration tth; β describes coefficient, which depends on M, in which $\beta = M_i^t / M_{\max}^t$, $r_1 = rand(0, 1)$, b represents a parameter, and $\overrightarrow{X_{Sun}^t}$ indicates the current location of the Sun in the iteration tth.

For Local search

$$\overrightarrow{X_i^{t+1}} = \overrightarrow{X_i^t} + c \times r_1 \times \left(r_2 \times \overrightarrow{X_{Sun}^t} - \overrightarrow{X_i^t}\right) \tag{5}$$

where $c = c_0 - t/T$, and r_2 is presented by Gauss distribution function.

In which, two phases, i.e. global and local search, are controlled by the $R_{\min}$ coefficient as follows:

$$R_{\min} = \|up - lb\| / R_0. \tag{6}$$

3 A 24-Bar 3D Truss Structure

A space truss with 24 members, which are linked with 13 nodes, is discussed in this section. It is a star-shaped steel truss. The space truss is shown in Fig. 1. The 3D truss has 24 members. Table 1 shows the nodal coordinates.

The purpose of this example is to demonstrate the effectiveness of the POA algorithm to deal with a complex system. Specifically, the 3D truss with 24 bars is evaluated in detail for damage process via three different scenarios. With 13 nodes, the total number of DOFs is $13 \times 3 = 39$. The boundary conditions are applied at points 1 to 6 to reduce the number of DOFs to 21.

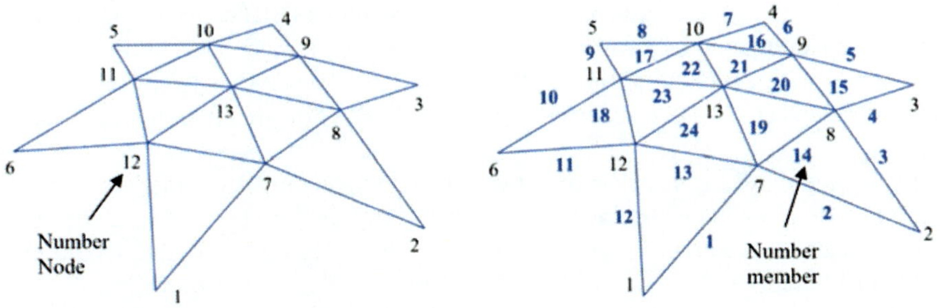

Fig. 1 The 3D truss structure (24 members)

Table 1 Nodal coordinates of 3D truss structure

Number node	X (cm)	Y (cm)	Z (cm)
1	0.0000	−50.0000	0.0000
2	43.3013	−25.0000	0.0000
3	43.3013	25.0000	0.0000
4	0.0000	50.0000	0.0000
5	−43.3013	25.0000	0.0000
6	−43.3013	−25.0000	0.0000
7	12.5000	−21.6506	6.2160
8	25.0000	0.0000	6.2160
9	12.5000	21.6506	6.2160
10	−12.5000	21.6506	6.2160
11	−25.0000	0.0000	6.2160
12	−12.5000	−21.6506	6.2160
13	0.0000	0.0000	8.2160

POA is employed for SHM based on natural frequencies. An objective function is defined as the difference in frequencies between FE model and true measurements. The assessment of SHM uses the change in the stiffness of the structure. Thereby, severity and location of damage will be predicted. To check the accuracy of this method, we also propose three damage scenarios. The scenarios are shown in Fig. 2 and Table 2. At the same time, two modern metaheuristics, namely GWO [8] and AOA [10], are also adopted to compare the performance of each algorithm in this problem.

For a fair comparison, all algorithms have the same initial conditions, namely population size of 30 individuals and maximum number of iterations of 500, for the three mentioned scenarios in Table 2. The obtained results are stored in a set of data after 30 runs to find out the best solution as shown in Figs. 3, 4, 5, 6, 7 and 8.

We can see that all optimizers are roughly identifying the potentially damaged location(s) in all cases. However, convergence curves show that the performance of the algorithms is different. Indeed, for the 1st scenario, the simplest problem, although the damage detection results of three algorithms satisfied the requirements for the measurement as shown in Fig. 4, POA shows promising results compared with other optimization algorithms. The first case with single damage results at the 200th iteration of POA are superior to AOA and GWO at the 500th iteration (see Fig. 3).

In the second scenario, and again, POA is more outstanding than other optimizers at the first iteration. The accuracy of POA at the 10th iteration providing the value of the objective function is more effective than other algorithms at the maximum number of iterations.

On the other hand, there is a significant difference in the results identifying the severity of damage using different algorithms for the last scenario. For this case, the complexity of the problem led to more challenges for all algorithms, and hence,

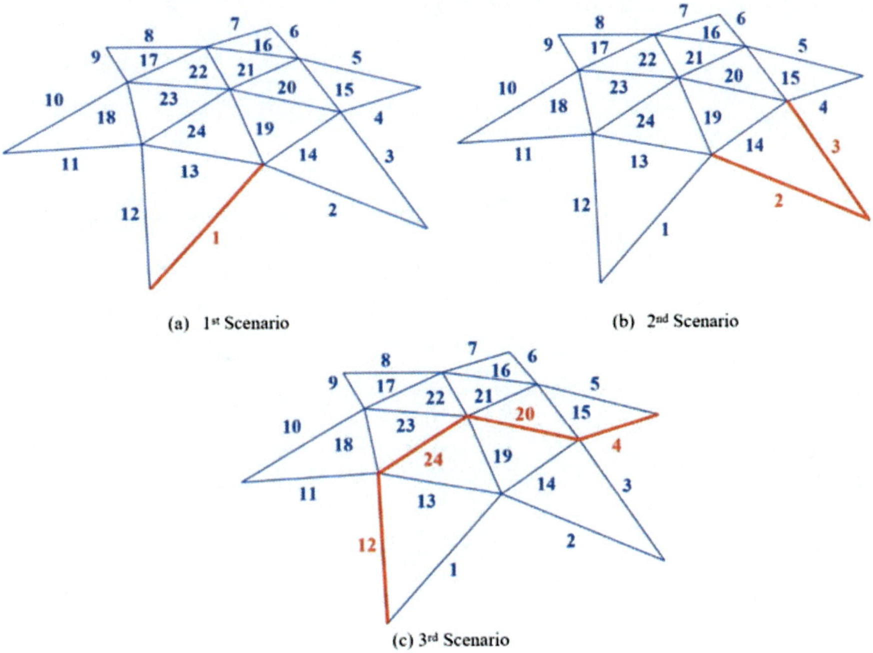

Fig. 2 The damage scenarios

Table 2 The damage scenarios of the 3D truss (24-member)

Scenarios	Damage bar(s)	Reduction rate (R) in the stiffness (E)
The 1st scenario	Bar 1	R = 50%
The 2nd scenario	Bar 2	R = 46%
	Bar 3	R = 35%
The 3rd scenario	Bar 4	R = 10%
	Bar 12	R = 15%
	Bar 20	R = 20%
	Bar 24	R = 25%

the 3rd scenario can be used to assess the performance of the algorithm thoroughly. Meanwhile, AOA made some errors in locating damaged site, such as bars 2, 5 and 13, that led to incorrect prediction for bar 4. At the same time, it affected the results in bars 20 and 24 (see Fig. 8). In contrast, GWO and POA show great performance in terms of exploration and exploitation of the optimum. Both optimizers predict the severity and location of damage accurately. Nevertheless, POA shows a higher performance than GWO in damage detection (see Fig. 7).

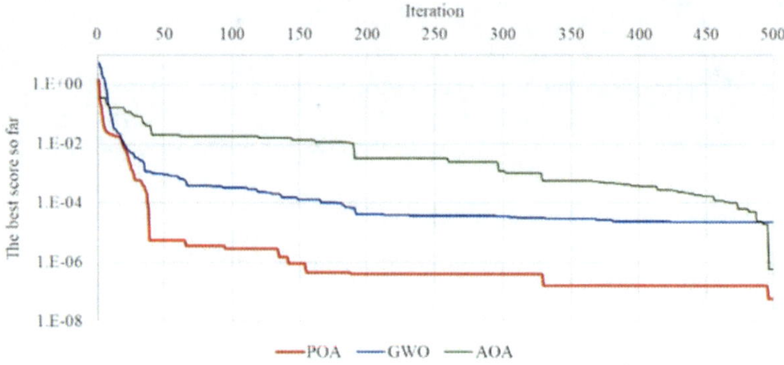

Fig. 3 Convergence curve of algorithms for the 1st scenario

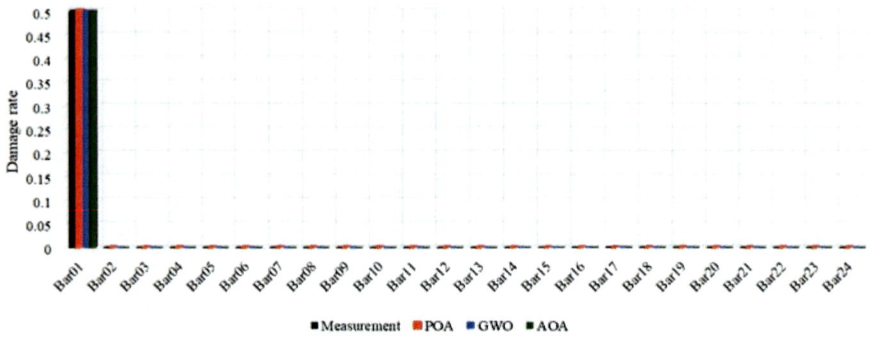

Fig. 4 The damage detection result versus the measurement for 1st scenario

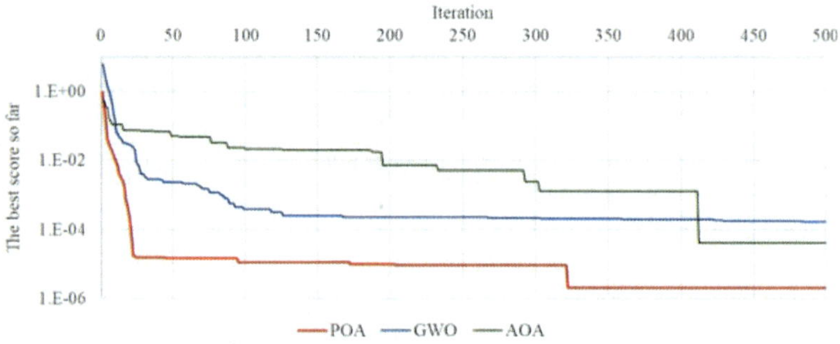

Fig. 5 Convergence curve of algorithms for the 2nd scenario

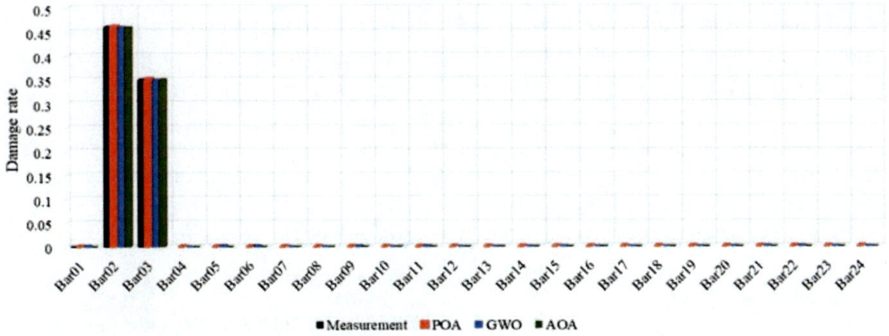

Fig. 6 The damage detection result versus the measurement for the 2nd scenario

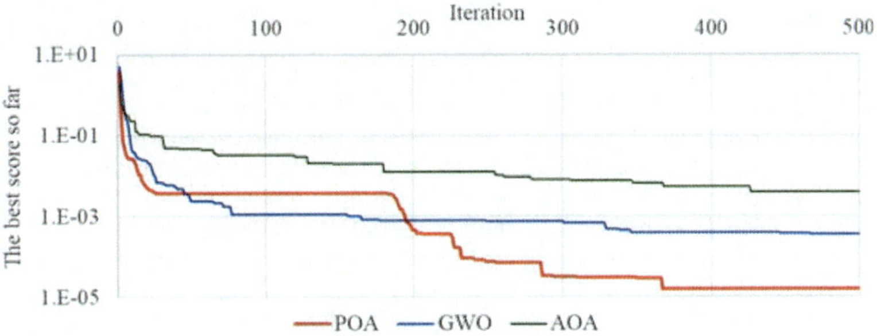

Fig. 7 Convergence curve of algorithms for the 3rd scenario

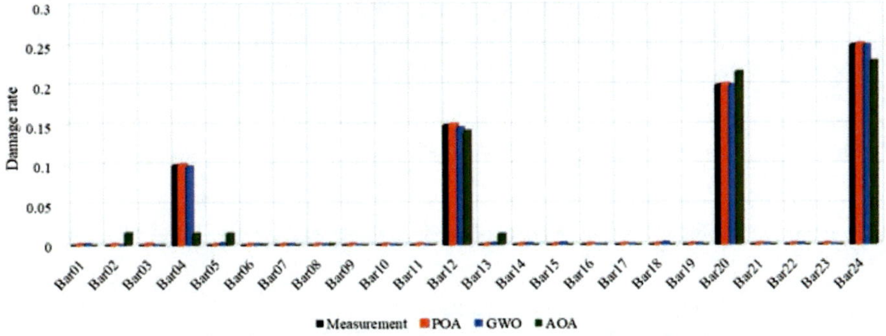

Fig. 8 The damage detection result versus the measurement for the 3rd scenario

4 Conclusion

In this study, POA, GWO and AOA are used for damage detection in 3D truss structure. A set of scenarios, from simple to complex, is proposed to assess the performance of the optimisation algorithms. The obtained results are compared thoroughly.

In most cases, POA shows the highest performance in damage detection scenarios. The obtained results prove that POA provides not only good value of the objective function, but also predicts the location and severity of damage accurately.

Acknowledgements The authors acknowledge the financial support of VLIR-UOS TEAM Project, VN2018TEA479A103, 'Damage assessment tools for Structural Health Monitoring of Vietnamese infrastructures' funded by the Flemish Government.

vliruos SHARING MINDS, CHANGING LIVES

With the support of THE BELGIAN DEVELOPMENT COOPERATION .be

References

1. Minh H-L, Sang-To T, Danh T-T, Phu N-N, Wahab MA, Cuong-Le T (2022) A two-step approach for damage detection in a real 3D tower using the reduced-order finite element model updating and atom search algorithm (ASO). In: Proceedings of the 2nd international conference on structural damage modelling and assessment. Springer, pp 13–26
2. Minh H-L, Khatir S, Wahab MA, Cuong-Le T (2021) An Enhancing Particle Swarm Optimization Algorithm (EHVPSO) for damage identification in 3D transmission tower. Eng Struct 242:112412
3. Sang-To T, Hoang-Le M, Wahab MA, Cuong-Le T (2022) Predicting damaged truss using meta-heuristic optimization algorithm. In: Recent advances in structural health monitoring and engineering structures. Springer
4. Sang-To T, Hoang-Le M, Wahab MA, Cuong-Le T (2022) An efficient Planet Optimization Algorithm for solving engineering problems. Sci Rep 12(1):1–18
5. To TS, Le MH, Danh T-T, Khatir S, Wahab MA, Le TC (2022) Combination of intermittent search strategy and an improve particle swarm optimization algorithm (IPSO) for model updating. Frattura ed Integrita Strutturale-Fract Struct Integrity 16(59):141–152
6. Cuong-Le T, Minh H-L, Khatir S, Wahab MA, Tran MT, Mirjalili S (2021) A novel version of Cuckoo search algorithm for solving optimization problems. Expert Syst Appl 186:115669
7. Minh H-L, Khatir S, Rao RV, Wahab MA, Cuong-Le T (2021) A variable velocity strategy particle swarm optimization algorithm (VVS-PSO) for damage assessment in structures. Eng Comput 1–30
8. Mirjalili S, Mirjalili SM, Lewis A (2014) Grey wolf optimizer. Adv Eng Softw 69:46–61
9. Minh H-L, Sang-To T, Wahab MA, Cuong-Le T (2022) A new metaheuristic optimization based on K-means clustering algorithm and its application for structural damage identification in a complex 3D concrete structure. Knowl-Based Syst 109189
10. Abualigah L, Diabat A, Mirjalili S, Elaziz MA, Gandomi AH (2021) The arithmetic optimization algorithm. Comput Methods Appl Mech Eng 376:113609
11. Sang-To T, Hoang-Le M, Khatir S, Mirjalili S, Wahab MA, Cuong-Le T (2021) Forecasting of excavation problems for high-rise building in Vietnam using planet optimization algorithm. Sci Rep 11(1):1–10

Effect of the Incident Wave Angle on the Hydrodynamic Performance of a Land-Based OWC Device

Ayrton Alfonso Medina Rodríguez, Gregorio Posada Vanegas, Beatriz Edith Vega Serratos, Alejandro Martínez Flores, Edgar Gerardo Mendoza Baldwin, Jesús María Blanco Ilzarbe, and Rodolfo Silva Casarín

Abstract The majority of experiments on fixed Oscillating Water Column (OWC) systems assume that water waves impact perpendicularly on the front wall of the device. However, this seldom occurs in practice due to wave transformation, which occurs when waves interact with shifting bottom profiles resulting in wave reflection, refraction and shoaling. The wave angle of incidence is of paramount relevance because it can alter the performance of the OWC device, particularly the natural period at which the device resonates. Therefore, this work investigates the interaction of directional waves with a fixed land-based OWC device. Theoretical and experimental techniques to study the effect of wave direction on the device hydrodynamic performance are described. The mathematical problem for the theoretical approaches is formulated using two-dimensional linear wave theory. The conventional eigenfunction expansion method (EEM) and the Boundary Element Method (BEM) are used to solve the governing equation together with the boundary conditions. Then, a series of experimental tests under regular wave conditions were carried out in a directional wave basin to compare and validate the theoretical results. The effects of wave angle of incidence on hydrodynamic efficiency are examined. Analytical and numerical predictions of the resonance frequency for different wave angles of incidence were found to be in good agreement when compared with experimental tests. Findings reveal that the resonant frequency of the system increases exponen-

A. A. Medina Rodríguez (✉) · A. Martínez Flores · E. G. Mendoza Baldwin · R. Silva Casarín
Institute of Engineering, National Autonomous University of Mexico, Circuito Escolar, CP 04510 Mexico City, Mexico
e-mail: ayrtonamedinar@gmail.com

G. Posada Vanegas · B. E. Vega Serratos
EPOMEX Institute, Autonomous University of Campeche, Av. Héroe de Nacozari 480, CP 24079 Campeche, Mexico

J. M. Blanco Ilzarbe
Department of Energy Engineering, Faculty of Engineering in Bilbao, UPV/EHU, Plaza Ingeniero Torres Quevedo, 1, CP 48013 Bilbao, Spain

M. Abdel Wahab (ed.), *Proceedings of the 5th International Conference on Numerical Modelling in Engineering*, Lecture Notes in Civil Engineering 311,
https://doi.org/10.1007/978-981-19-8429-7_10

tially when the incident wave angle increases, a trend that is more visible for wave angles beyond 15°. Results indicate that analytical and numerical techniques can be employed as design tools to estimate the natural frequency of the system when it interacts with oblique regular waves.

Keywords Oscillating water column · Oblique waves · Hydrodynamic efficiency · Resonant frequency · Wave energy

1 Introduction

In recent years, a wide range of technologies has been developed which, as they are deployed, will contribute to meeting the energy demands of the world's rapidly growing population, while reducing the consequences associated with the use of fossil fuels. This renewable energy source has attracted the curiosity of numerous inventors and researchers, who have proposed a diverse variety of wave extraction devices, with over a thousand wave energy patents to date [1]. Among the several suggested technologies, the OWC device has been shown to be one of the most promising technologies for harvesting ocean wave energy.

The OWC device is a one-of-a-kind wave energy converter (WEC) system, as it has just two major components: a partially submerged collecting chamber and a power take-off (PTO) mechanism that is placed above sea level. The basic working concept is that the water column inside the collecting chamber oscillates vertically, owing to the incident wave, forcing the confined air volume up and down through a turbine connected to a generator [2]. The OWC is available in a variety of configurations (shore-mounted, sea-bed, and floating systems), allowing it to be deployed onshore and offshore.

Despite these advantages, one of the most significant challenges with this type of WEC is its sophisticated hydrodynamic mechanism, which involves complicated diffraction and radiation wave mechanisms [1]. Theoretical hydrodynamics has characterised most of the work on OWC systems published in the last three decades, particularly the wave-structure interaction, under the assumption that wavefronts travel normally towards the device [3–7]. However, under real-world conditions, the wave direction is not necessarily perpendicular to the OWC chamber's transverse axis [8].

Previous research has used analytical, experimental and numerical methodologies to study the effect of wave direction on the hydrodynamic performance of fixed OWC devices [8–15]. The influence of wave direction on the performance of the OWC chamber was investigated by [8]. They conducted experiments in a three-dimensional wave basin for distinct wave directions that include the effect of the turbine through a hole in the physical model chamber. They found that, for wave angles perpendicular or close to perpendicular to the OWC front wall, the wave direction had a negligible effect on water surface elevation inside the chamber. In turn, as the incidence moves away from perpendicularity, the efficiency of the OWC decreases.

Ashlin et al. [11] investigated the performance of an array of OWC devices connected to an offshore detached breakwater that was subjected to oblique wave incidence. They found that when the wave angle decreases, the system's performance decreases because the wavefront impacts the system at varying moments, while the fundamental frequency of the system remains constant. Medina Rodríguez et al. [13] used the matched EEM and BEM to investigate the interaction of oblique water waves with a land-fixed OWC device. They concluded that increasing the angle of the incident wave results in a wider hydrodynamic efficiency band as well as a higher wave frequency at which resonance occurs.

Several studies on the hydrodynamic performance of land-based OWC systems have been conducted, however, the majority of research has focused on the two-dimensional interaction of water waves with land-based OWC devices. OWC devices are designed to serve as terminator devices (capturing the greatest energy by having their major axis perpendicular to the predominant wave direction), although this seldom occurs in practice. For example, incident water waves impacting a particular Wave Energy Plant of Ref. [16] are often oblique [17].

2 Aims and Methodology

In this work, the analytical and numerical techniques employed by [14] are compared with the experimental results obtained for a scaled thick-front wall OWC device, corresponding to one chamber of the Mutriku Wave Power Plant (MWPP and located in the Basque Country, Spain), in a directional wave basin. To supplement prior studies on land-based OWCs, the goals of this work are to focus on the suitability of theoretical and numerical approaches (EEM and BEM, respectively) to estimate the natural frequency of a fixed OWC system and the influence of incident wave direction on the hydrodynamic efficiency.

3 Mathematical Approach

The OWC device and the Cartesian coordinate system are shown in Fig. 1. The water depth is defined as h, and the origin is on the undisturbed free surface. The waves come from the x direction towards the land-based OWC device and form an angle θ with the x-axis. The OWC comprises a partially submerged front wall at $x = b$, with draft a and thickness w, and a back vertical wall located at $x = 0$. A Wells-type turbine is assumed to connect the confined air within the chamber to the atmosphere. Because we are just considering the two-dimensional scenario, the OWC is assumed to be infinitely long and parallel to the incoming wave crest. The boundaries are defined as

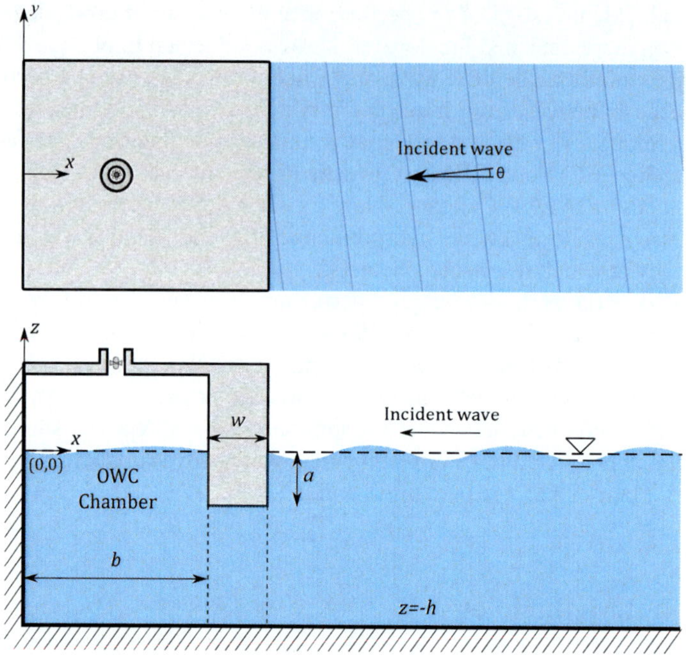

Fig. 1 Diagram of the land-based OWC device-directional waves interaction

- The front barrier by $B_b = \{(x, z) : (x = b, -a \leq z \leq 0) \cup (b < x < f_w.z = -h_a) \cup (x = f_w, -a \leq z \leq 0)\}$ with $f_w = b + w$.
- The back wall by $B_w = \{(x, z) : x = 0, -h < z < 0\}$.
- The free surface within the chamber by $F_i = \{(x, z) : 0 \leq x \leq b, z = 0\}$.
- The free surface outside the chamber by $F_f = \{(x, z) : f_w \leq x \leq \infty, z = 0\}$.
- The horizontal floor by $B_d = \{(x, z) : (0 < x < \infty, z = -h)\}$.
- The height of the aperture beneath the front barrier by $B_g = \{(x, z) : x = b, -h \leq z \leq -a\}$.

The water is considered to be incompressible and inviscid with an irrotational wave motion. The linearized wave theory is considered to effectively describe the wave motion and the surface tension effects are neglected; thus potential theory is used. A velocity potential then defined as

$$\Phi(x, z, t) = \mathrm{Re}\{\phi(x, z)\mathrm{e}^{-\mathrm{i}\omega t + \mathrm{i}\beta y}\}, \tag{1}$$

where ω is the angular frequency of the simple harmonic flow, the real part of a complex expression is denoted by Re{ }, $\beta = k \sin\theta$, k is the wavenumber of the plane wave and t is the time. The spatial velocity potential ϕ is governed by the Helmholtz equation under these assumptions, which is expressed as

$$\left(\frac{\partial^2}{\partial x^2} + \frac{\partial^2}{\partial z^2} - \beta^2\right)\phi = 0, \tag{2}$$

In addition to the impermeable boundary conditions imposed at the horizontal floor, front barrier, and back wall described by

$$\frac{\partial \phi}{\partial n} = 0 \quad \text{on} \quad B_d, \quad B_b \text{ and } B_w, \tag{3}$$

respectively.

On the free surface, the linear boundary conditions are given by

$$\frac{\partial \phi}{\partial z} - K\phi = \begin{cases} \frac{\mathrm{i}\omega p}{\rho g} & \text{on} \quad z = 0, \quad 0 < x < b, \\ 0 & \text{on} \quad z = 0, \quad f_w < x < \infty, \end{cases} \tag{4}$$

respectively, with $K = \omega^2/g$, g is the gravitational acceleration, p represents the harmonic pressure distribution in the free surface within the OWC chamber, while ρ is the seawater density.

Following the procedure of [18], the velocity potential is decomposed into a scattered potential ϕ^S and a radiated potential ϕ^R as follows

$$\phi(x, z) = \phi^S + \frac{\mathrm{i}\omega p}{\rho g}\phi^R, \tag{5}$$

where ϕ^S and ϕ^R satisfy Eqs. (2)–(4), with $p = 0$ inside the OWC chamber for ϕ^S; while for ϕ^R Eq. (4) is replaced by

$$\frac{\partial \phi^R}{\partial z} - K\phi^R = 1 \quad \text{on} \quad z = 0, \quad 0 < x < b. \tag{6}$$

At the right end of the domain ($x \to +\infty$), the Sommerfeld radiation condition is applied

$$\frac{\partial \phi^{D,R}}{\partial x} - \mathrm{i}k\cos\theta\phi^{D,R} = 0 \quad \text{as} \quad x \to +\infty, \tag{7}$$

where ϕ^D is the diffracted potential and $\phi^S = \phi^D + \phi^I$ with ϕ^I being the incident potential, while k is the real root of the dispersion equation given by

$$\omega^2 = gk\tanh kh. \tag{8}$$

Similar to the decomposition of ϕ, the induced volume flux across the internal free surface q is separated as follows

$$q = \int_{F_i} \frac{\partial \phi}{\partial z} dx = q^S + \frac{\mathrm{i}\omega p}{\rho g} q^R. \tag{9}$$

The volume flux continuity between the internal free surface and the gap below the front barrier is ensured by

$$q^{S,R} = \int_{F_i} \frac{\partial \phi^{S,R}}{\partial z} dx = \int_{B_g} \frac{\partial \phi^{S,R}}{\partial x} dz. \tag{10}$$

4 Method of Solution

4.1 Eigenfunction Expansion Method

The solution approach based on the matched EEM, as described by [14], is used in this work. This method was chosen because of the simple OWC geometry considered in this work, with a separable governing Helmholtz equation (2) and known horizontal and vertical eigenfunctions. The domain is divided into three sections for this purpose, see Fig. 2. First, the velocity potentials in each of the three regions are expanded in terms of the corresponding eigenfunctions. Then, with the aid of the continuity equations of pressure and horizontal velocity on the lateral sides of Region 2 defined as

$$\phi_- = \phi_+ \quad \text{and} \quad \frac{\partial \phi}{\partial x}_- = \frac{\partial \phi}{\partial x}_+ \quad \text{on} \quad x = b \text{ and } f_w \quad \text{with} \quad -h \leq z \leq -a. \tag{11}$$

Then by utilising the orthonormality of the vertical eigenfunctions, a linear system of algebraic equations to solve for the unknowns is obtained.

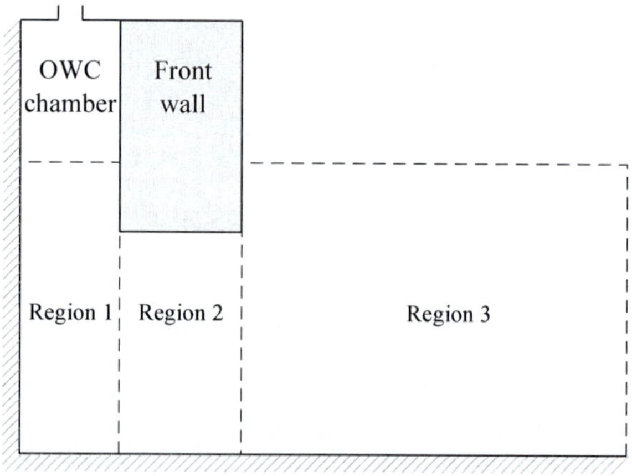

Fig. 2 Separation of the domain

4.2 Boundary Element Method

The Helmholtz equation (2) is solved using the BEM and the subdomain approach as in [14]. This method is an efficient numerical approach that provides reasonably accurate results [6, 13, 19]. The Helmholtz expression (2) is represented by its boundary integral equation as

$$\alpha(X)\phi(X) + \int_{\Gamma} \phi(Y)\frac{\partial \psi(X, Y)}{\partial n_Y} d\Gamma_Y = \int_{\Gamma} \psi(X, Y)\frac{\partial \phi(Y)}{\partial n_Y} d\Gamma_Y, \tag{12}$$

with ϕ and $\partial\phi/\partial n$ being, respectively, the unknown velocity potential and the normal derivative of ϕ with respect to the field point $Y\left(\hat{\varsigma}, \hat{\eta}\right)$ on the boundary Γ; $X\,(x, z)$ being the source point inside the domain Ω; ψ and $\partial\psi/\partial n$ being, respectively, the fundamental solution of Helmholtz equation and its normal derivative at point Y in Γ; and $\alpha = \tau/2\pi$, with τ being the angle in radians between points X and Y [20].

For Helmholtz equation, the fundamental solution is defined as

$$\psi = \frac{K_0\,(kr\sin\theta)}{2\pi}, \tag{13}$$

where K_0 is the modified Bessel function of the second kind and zero order which satisfies Eq. (2) and $r = \sqrt{\left(x - \hat{\varsigma}\right)^2 + \left(z - \hat{\eta}\right)^2}$ is the distance between points X and Y.

Finally, for normal incidence propagation ($\theta = 0$), Laplace equation governs the velocity potential ϕ and is defined as

$$\left(\frac{\partial^2}{\partial x^2} + \frac{\partial^2}{\partial z^2}\right)\phi = 0, \tag{14}$$

while its fundamental solution as

$$\psi = \frac{1}{2\pi}\ln r. \tag{15}$$

5 Experimental Investigation

5.1 Wave Basin

Experiments were carried out in a wave basin housed in the EPOMEX institute located at the Autonomous University of Campeche, Mexico. The wave basin is 15 m long, 9 m wide and 0.8 m deep. At one end of the basin, a snake-type wavemaker with

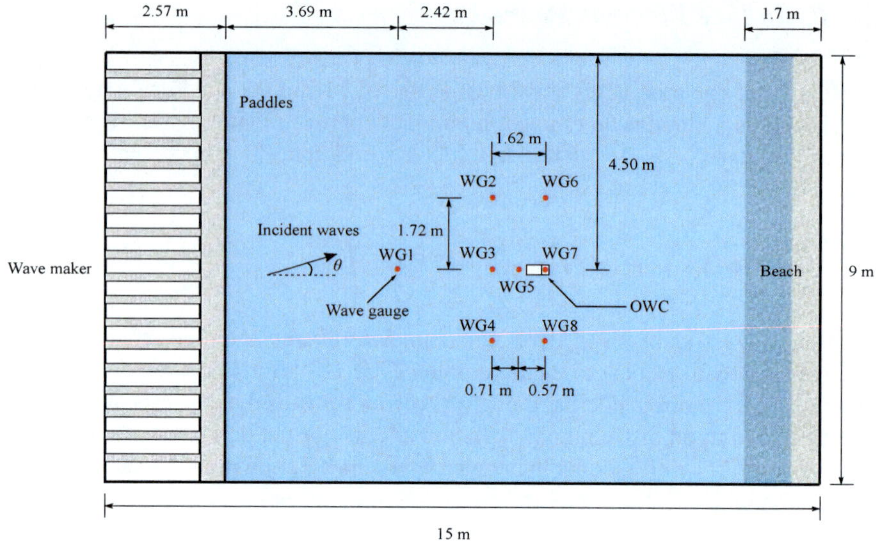

Fig. 3 Schematic representation of the wave basin and the experimental set-up

18 piston paddles (50 cm wide each) is controlled by an integrated and synchronised data acquisition system. At the other end, an extinction system acts as a passive wave absorber. This artificial beach is made of gravel with a medium value of the gravel size distribution D_{50} =5.08 cm and an average weight of 54 g. Figure 3 shows a schematic diagram of the directional wave basin along with the positions of eight wave gauges and the scaled OWC model tested.

5.2 *Test Model*

The OWC model was built at a 1:20 scale following Froude similarity criteria. Froude scaling was employed to represent an analogous OWC system that is geometrically equivalent. The size of the wave basin, as well as the wave conditions under consideration, defined this scale ratio. It should be noted that when applying Froude scaling, other force-to-force ratios might be distorted [21–23]. In this regard, viscous forces are generally minimal over short distances, and surface tension effects are typically small for models with water depths and wave periods greater than 20 mm and 0.35 s, respectively [24].

Figure 4a shows the dimensional details of the OWC model. The dimensions are geometrically equivalent to one OWC chamber of the MWPP [14]. The dimensions of the model are defined in Table 1. The OWC model was built using acrylic sheets, 12 mm thick, that were cut with a laser system. Throughout the studies, the water depth, h = 400 mm, was kept constant. Extra weights (8 kg), made of lead, were used

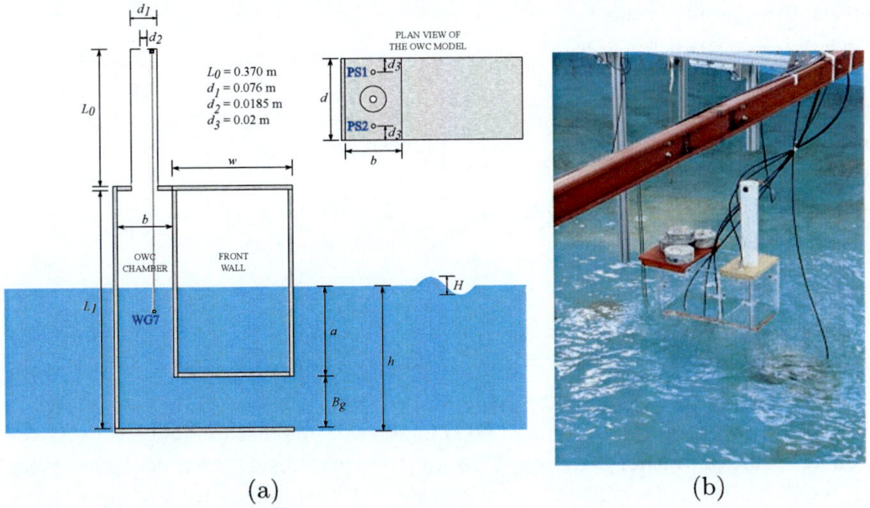

Fig. 4 Scaled OWC model and experimental set-up in the wave basin. **a** Cross section: details of the OWC device. **b** The experimental set-up of the OWC model

Table 1 OWC model geometric characteristics

Parameter	Value (m)
Length (b)	0.155
Width (d)	0.255
Height (L_1)	0.655
Draft (a)	0.260
Front barrier thickness (w)	0.333
Gap height (B_g)	0.128

to assure the stability of the OWC device. The distance between the OWC model and the artificial beach is 3.0 m, aiming to ensure enough measurement time, preventing wave reflection, as depicted in Figs. 3 and 4b. The PTO unit acting on the OWC system in this investigation was an impulse turbine. A circular orifice was used to replicate the PTO, which applied the equivalent resistance of a self-rectifying impulse turbine [23]. The air hole area to waterplane area ratio (χ) of the OWC chamber was 0.68%, as in [25], and it was kept constant throughout the experiments.

5.3 Instrumentation

The water-free surface oscillations outside the OWC model were recorded by seven resistance-type wave gauges (0.01–0.70 m VTI, WG-1CH-E, Spain) installed within

Table 2 Experimental conditions

Parameter	Values
Water depth (h)	0.4 m
Incident wave height (H)	0.02–0.10 m with intervals of 0.02 m
Wave period (T)	1.0–3.0 s with intervals of 0.2 s
Wavelength (λ)	1.464, 1.936, 2.393, 2.836, 3.269, 3.695, 4.115, 4.532, 4.945, 5.356 and 5.765 m
Incident wave direction (θ)	0°–30° with intervals of 15°

the wave basin. Wave gauges 1–5 were installed in front of the OWC model, while WG6 and WG8 were mounted at the model's lateral sides, as previously shown in Fig. 3. One resistance wave gauge (WG7) and two air pressure gauge sensors (0–0.3 bar Acculevel, Keller, Virginia, USA) were placed inside the model's chamber, Fig. 4a. All of the sensors had a sampling rate of 100 Hz. The pressure transducers, on the other hand, have a measurement range of 0.3 bar. The power absorbed by the system was calculated using the differential air pressure obtained by PS1 and PS2, as well as the water-free surface elevation of the chamber, measured by WG7. The measurements were analysed within the steady-state time frame.

5.4 Experimental Wave Conditions

The OWC device was tested under regular wave conditions. Experimental conditions are summarised in Table 2. The performance of the OWC model was studied by varying the wave height, wave period and incident wave angles.

6 Hydrodynamic Efficiency

To evaluate the system performance, the hydrodynamic efficiency was defined as

$$\epsilon = \frac{P_{out}}{P_{in}}, \tag{16}$$

where P_{in} is the available power over one wave period of a monochromatic wave given by

$$P_{in} = Ec_g \tag{17}$$

with E being the total energy per wave period and c_g the group velocity given by

$$E = \frac{1}{2}\rho g d \left(\frac{H}{2}\right)^2, \tag{18}$$

$$c_g = \frac{1}{2}\frac{\omega}{k}\left(1 + \frac{2kh}{\sinh(2kh)}\right), \tag{19}$$

respectively.

The average power absorbed by regular waves P_{out} can be calculated by integrating the instantaneous free surface oscillation within the chamber moving with a velocity (V_{fs}) under the air pressure (ΔP) as follows

$$P_{out} = \frac{1}{t_{fin} - t_{ini}} \int_{t_{ini}}^{t_{fin}} \Delta P S_{wpa} V_{fs} dt, \tag{20}$$

where t_{ini} and t_{fin} are the starting and final times of the measurements in the steady state region; $S_{wpa} = b \times d$ is the water plane area of the OWC chamber and t is the time. The parameter V_{fs} can be determined by taking the first-time derivative of the third-order approximation to the free surface elevation within the OWC chamber [26] as

$$V_{fs} = \frac{2\eta_{j+1} + 3\eta_j - 6\eta_{j-1} + \eta_{j-2}}{6\Delta t}, \tag{21}$$

where η_j is the free surface elevation at time t_j, j is the current time value and Δt is the sampling interval. Furthermore, the air pressure ΔP within the chamber is calculated by averaging the data collected from the two pressure gauges, PS1 and PS2, at each time instant.

7 Results

In this section, the theoretical and experimental results based on the hydrodynamic efficiency of a particular land-based OWC device are shown. The chosen values, based on a single chamber of the MWPP, for performing the two-dimensional numerical calculations were $h = 8.0$ m, $a = 5.20$ m, $b = 3.10$ m and $w = 6.65$ m with a wave period T in the range of $2.80 \le T \le 30$ s. Thus, the hydrodynamic efficiency ε for different incidence wave angles and wave heights are presented and discussed.

Figure 5a, b shows the hydrodynamic efficiency ε for different wave angles of incidence versus h/λ obtained by the matched EEM and the BEM. Figure 5a, b demonstrate that the semi-analytical and numerical findings obtained by the EEM and the BEM are in excellent agreement. It is seen in Fig. 5a, b that the hydrodynamic efficiency curves for $\theta = 0°$ and $15°$ are almost superimposed; however, with the aid of Fig. 6, which shows the results of ε obtained by EEM as a function of h/λ and θ, it is demonstrated that the effect of wave direction on the hydrodynamic efficiency is significant for θ beyond $15°$. It is also observed that as θ increases, the value of the

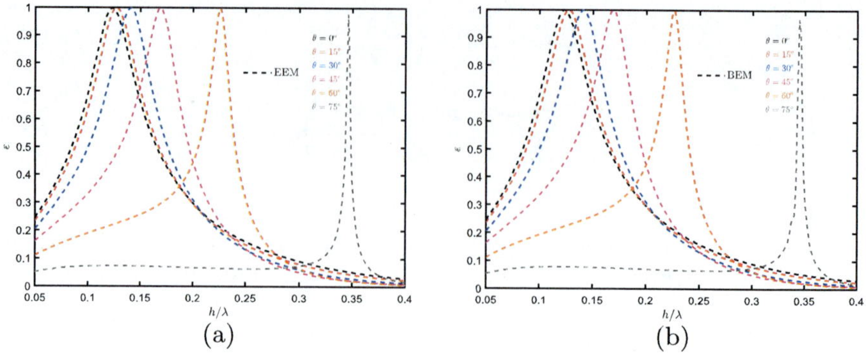

Fig. 5 Hydrodynamic efficiency ε versus h/λ for different wave direction (θ) with $b/h = 0.3875$, $a/h = 0.65$ and $w/h = 0.83125$. **a** Analytical results obtained by the matched EEM. **b** Numerical results obtained by the BEM

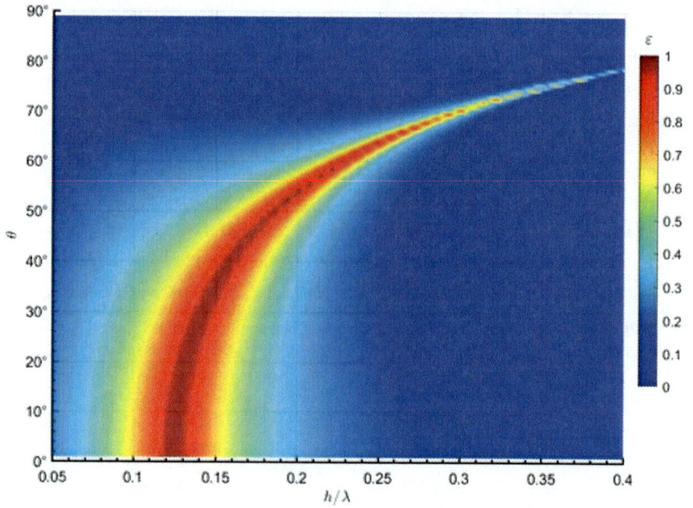

Fig. 6 Analytical results of the hydrodynamic efficiency ε as a function of h/λ and θ with $b/h = 0.3875$, $a/h = 0.65$ and $w/h = 0.83125$

non-dimensional frequency h/λ at which resonance occurs also increases, while the hydrodynamic efficiency bandwidth is significantly reduced. From Fig. 5a, b, the values of $h/\lambda(=0.126,\ 0.129,\ 0.143,\ 0.170,\ 0.228$ and $0.348)$ where the maximum value of ε occurs, correspond to wave periods $T(=7.85,\ 7.70,\ 7.08,\ 6.18,\ 5.02$ and 3.89 s), respectively. To make better use of the available wave energy, in practice, this feature of the resonance frequency-changing due to the direction of the waves will be a key argument to decide where to instal a land-based OWC device.

From Fig. 7a–e, the experimental results for the hydrodynamic efficiency ε versus h/λ for different wave height to water depth ratios and wave angles of incidence are

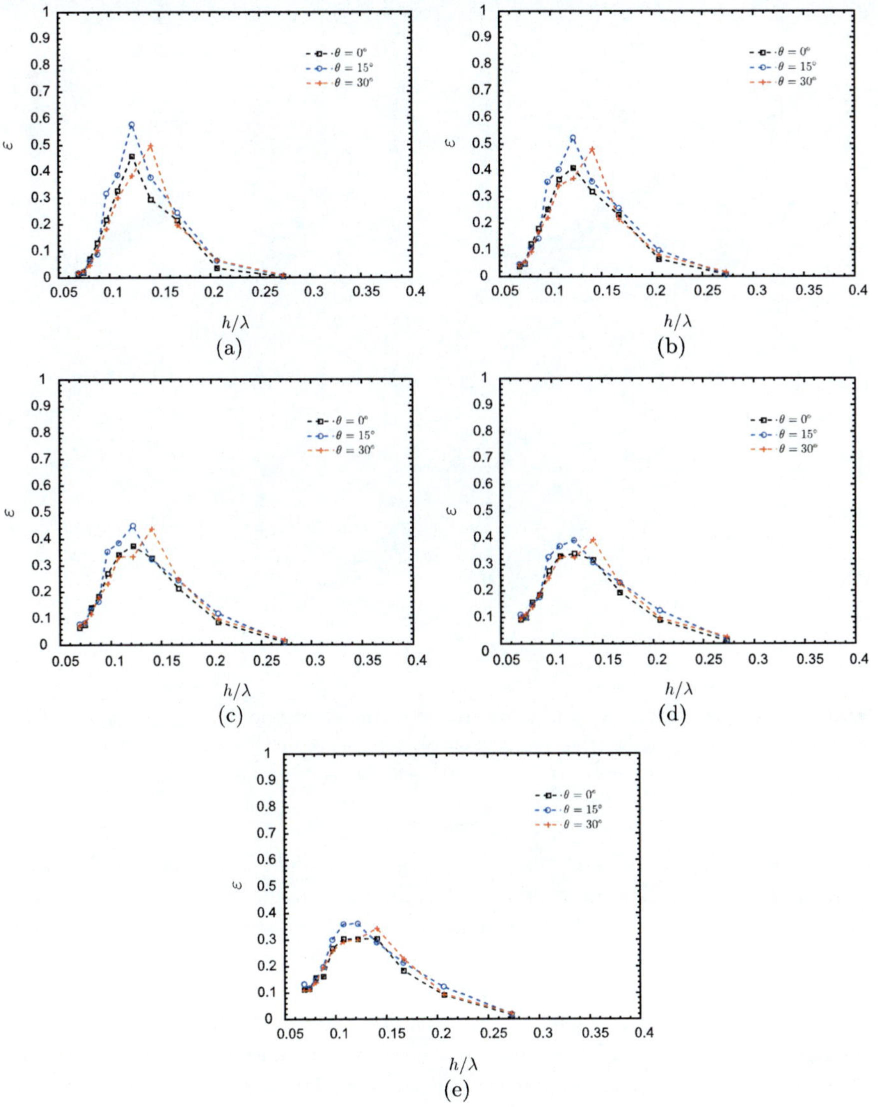

Fig. 7 Hydrodynamic efficiency ε versus h/λ for different wave height to water depth ratios (H/h) and wave direction (θ). Experimental results for **a** $H/h = 0.05$. **b** $H/h = 0.10$. **c** $H/h = 0.15$. **d** $H/h = 0.20$. **e** $H/h = 0.25$

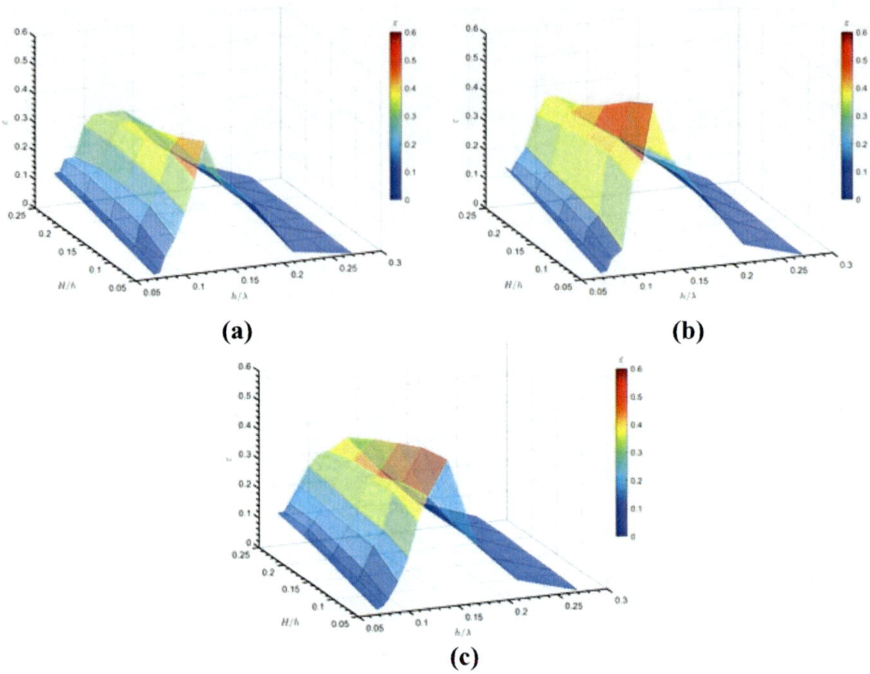

Fig. 8 Hydrodynamic efficiency ε as a function of the non-dimensional frequency h/λ and the wave height to water depth ratio H/h for the three different wave directions. **a** Experimental results for $\theta = 0°$. **b** Experimental results for $\theta = 15°$. **c** Experimental results for $\theta = 30°$

shown. In these figures, as was observed in Figs. 5a, b and 6, it is seen that the value of h/λ at which ε is maximum increases as the wave angle also increases. This is better observed for $\theta = 15°–30°$. This pattern is because shorter waves that reach the structure at an oblique angle can excite the system with lower energy dissipation. It also means that for bigger wave angles, the OWC chamber must interact with shorter wavelengths to resonate. From these figures, it is also observed that wave height does not influence the value of h/λ at which resonance occurs and that the peak values in hydrodynamic efficiency are higher for smaller wave height to water depth ratios.

The hydrodynamic efficiency ε as a function of the non-dimensional frequency h/λ and the wave height to water depth ratio H/h is shown in Fig. 8a–c. These figures, along with Fig. 7a–e, show that ε increases when H/h decreases, regardless of the wave direction, and that the hydrodynamic efficiency is significantly reduced for short wavelengths $h/\lambda > 0.20$. Moreover, it can be observed that the values of h/λ where ε is maximum remain unaltered for the H/h ratio.

Finally, Table 3 gives a comparison between the analytical, numerical and experimental techniques of the values of the non-dimensional frequency h/λ at which resonance occurs for different wave directions. The values computed from the theoretical and numerical tools are in accordance with those obtained in the experimental

Table 3 Comparison of theoretical, numerical and experimental h/λ values at which resonance occurs for different incident wave angles

	Wave direction					
	0°	15°	30°	45°	60°	75°
Method	h/λ					
EEM	0.126	0.129	0.143	0.170	0.228	0.348
BEM	0.124	0.129	0.143	0.170	0.227	0.346
Experimental	0.122	0.122	0.141	–	–	–

tests. This demonstrates how potential flow-based procedures, such as the matched EEM and BEM, can be used for the preliminary design of a land-based OWC chamber to resonate for a specific wave period of a particular location. Then, findings can be corroborated by experimental testing of the selected OWC model, under regular wave conditions.

8 Conclusions

Analytical and numerical methodologies, as well as experimental testing under regular waves, were used to evaluate the performance characteristics of a land-based OWC device exposed to diverse incident wave directions and wave heights. The findings of this study are summarised below:

- The matched EEM and BEM showed to be appropriate for accurately predicting the resonance frequency of a land-fixed OWC model, with a thick-front wall, when it interacts with regular waves at different angles of incidence. These two methods can be used for preliminary designs of OWC chambers based on the local wave climate.
- The resonant frequency of the system increases exponentially when the incident wave angle increases. This trend was more visible for wave angles beyond 15° with increments of at least 8% in both numerical and experimental techniques.
- Regarding the effect of the wave height, it was observed that this does not affect the value of the resonance frequency. However, the magnitude of the peak value of the hydrodynamic efficiency increases when wave height decreases.

Additional numerical and experimental tests with varied OWC chamber designs and incidence wave angles are required to confirm the applicability of the matched EEM and BEM for estimating the resonant frequency of land-fixed OWC devices. Finally, it is expected that the results of this study will lead to successful wave energy harvesting and motivate further investigation into the interaction of land-based OWC devices with directional waves.

Acknowledgements The present study was conducted within the framework of CEMIE-Océano (Mexican Centre for Innovation in Ocean Energy). Project FSE-2014-06-249795 financed by CONACYT-SENER- Sustentabilidad Energética. The authors would like to thank the Basque Government through the research group (IT1514-22) for the guidance provided.

Abbreviations

In this work, the following abbreviations are used:

BEM	Boundary element method
EEM	Eigenfunction expansion method
MWPP	Mutriku Wave Power Plant
OWC	Oscillating water column
PS	Pressure sensor
PTO	Power take-off
WEC	Wave energy converter
WG	Wave gauge

Nomenclature

a	Front wall draft
b	Chamber length
B_b	Thick front wall boundary
B_d	Horizontal bottom boundary
B_g	Gap length
B_w	Rigid vertical wall boundary
c_g	Group velocity
d	Chamber width
E	Total energy per wave period
F_f	External free surface
F_i	Internal free surface
g	Gravitational acceleration
h	Water depth
H	Wave height
k	Wave number
L_1	Model height
n	Normal unit vector
p	Spatial pressure distribution
P_{in}	Available power over one wave period
P_{out}	Average power absorbed from regular waves
q	Volume flux

q^R	Radiated volume flux
q^S	Scattered volume flux
r	Distance between X and Y
$S_{chamber}$	Water plane area of the OWC chamber
t	Time
T	Incident wave period
T_{ini}	Initial time in the steady state region
T_{fin}	Final time in the steady state region
V_{fs}	Instantaneous free surface velocity
w	Front wall thickness
x	Horizontal axis
X	Source point
Y	Field point
z	Vertical axis

Greek Letters

α	Internal angle parameter
Γ	Boundary
ε	Hydrodynamic efficiency
η	Free surface elevation
θ	Wave angle of incidence
λ	Wavelength
ρ	Density of water
ϕ	Spatial velocity potential
ϕ^D	Diffracted velocity potential
ϕ^I	Incident velocity potential
ϕ^R	Radiated velocity potential
ϕ^S	Scattered velocity potential
Φ	Time-dependent velocity potential
ψ	2D fundamental solution of Helmholtz equation
ω	Angular frequency

References

1. Falcão AFO (2010) Wave energy utilization: a review of the technologies. Renew Sustain Energy Rev 14(3):899–918
2. Falcão AFO, Henriques JCC (2016) Oscillating-water-column wave energy converters and air turbines: a review. Renew Energy 85:1391–1424
3. Evans D, Porter R (1995) Hydrodynamic characteristics of an oscillating water column device. Appl Ocean Res 17:155–164
4. Morris-Thomas MT, Irvin RJ, Thiagarajan KP (2006) An investigation into the hydrodynamic efficiency of an oscillating water column. J Offshore Mech Arct 129:273–278
5. Şentürk U, Özdamar A (2012) Wave energy extraction by an oscillating water column with a gap on the fully submerged front wall. Appl Ocean Res 37:174–182

6. Rezanejad K, Bhattacharjee J, Guedes Soares C (2013) Stepped sea bottom effects on the efficiency of nearshore oscillating water column device. Ocean Eng 70:25–38
7. Koley S, Trivedi K (2020) Mathematical modeling of oscillating water column wave energy converter devices over the undulated sea bed. Eng Anal Bound Elem 117:26–40
8. Jin J, Liu Z, Hyun BS, Hong K (2012) Effects of wave direction on performance of oscillating water column type wave energy convertor. In: Proceedings of the international offshore and polar engineering conference, pp 582–587
9. Martins-rivas H, Mei CC (2009) Wave power extraction from an oscillating water column along a straight coast. Ocean Eng 36(6):426–433
10. Nader JR, Zhu SP, Cooper P, Stappenbelt B (2012) A finite-element study of the efficiency of arrays of oscillating water column wave energy converters. Ocean Eng 43:72–81
11. John Ashlin S, Sannasiraj SA, Sundar V (2016) Hydrodynamic performance of an array of oscillating water column device exposed to oblique waves. In: Proceedings of the 12th international conference on hydrodynamics. Egmond aan Zee, The Netherlands
12. Malara G, Gomes RPF, Arena F, Henriques JCC, Gato LMC, Falcão AFO (2017) The influence of three-dimensional effects on the performance of U-type oscillating water column wave energy harvesters. Renew Energy 111:506–522
13. Medina Rodríguez AA, Martínez Flores A, Blanco Ilzarbe JM, Silva Casarín R (2021) Interaction of oblique waves with an oscillating water column device. Ocean Eng 228:108931
14. Medina Rodríguez AA, Silva Casarín R, Blanco Ilzarbe JM (2022) The influence of oblique waves on the hydrodynamic efficiency of an onshore OWC wave energy converter. Renew Energy 183:687–707
15. Medina Rodríguez AA, Posada Vanegas G, Silva Casarín R, Mendoza Baldwin EG, Vega Serratos BE, Puc Cutz FE, Mangas Che EA (2022) Experimental investigation of the hydrodynamic performance of land-fixed nearshore and onshore oscillating water column systems with a thick front wall. Energies 15:2364. https://doi.org/10.3390/en15072364
16. Torre-Enciso Y, Ortubia I, López De Aguileta LI, Marqués J (2009) Mutriku wave power plant: from the thinking out to the reality. In: Proceedings of the 8th European wave and tidal energy conference (EWTEC), Uppsala, Sweden, 7–10 September 2009
17. Ibarra-Berastegi G, Sáenz J, Ulazia A, Serras P, Esnaola G, Garcia-Soto C (2018) Electricity production, capacity factor, and plant efficiency index at the Mutriku wave farm (2014–2016). Ocean Eng 147:20–29
18. Evans DV (1982) Wave-power absorption by systems of oscillating surface pressure distributions. J Fluid Mech 114:481–499
19. Medina Rodríguez AA, Silva Casarín R, Blanco Ilzarbe JM (2021) A theoretical study of the hydrodynamic performance of an asymmetric fixed-detached OWC device. Water 13(19):2637
20. Katsikadelis J (2002) Boundary elements, theory and applications. Elsevier, Amsterdam, The Netherlands
21. Sarmento AJNA (1993) Model-test optimization of an OWC wave power plant. Int J Offshore Polar Eng 3
22. Weber J (2007) Representation of non-linear aero-thermodynamic effects during small scale physical modelling of oscillating water column wave energy converters. In: Proceedings of the European wave and tidal energy conference EWTEC, Porto, Portugal
23. Falcão AFO, Henriques JC (2014) Model-prototype similarity of oscillating-water-column wave energy converters. Int J Mar Energy 6:18–34
24. Hughes SA (1993) Physical models and laboratory techniques in coastal engineering. World Scientific
25. John Ashlin S, Sundar V, Sannasiraj S (2016) Effects of bottom profile of an oscillating water column device on its hydrodynamic characteristics. Renew Energy 96:341–353
26. López I, Pereiras B, Castro F, Iglesias G (2015) Performance of OWC wave energy converters: influence of turbine damping and tidal variability. Int J Energy Res 39:472–483

Data-Driven Kriging Model for Predicting Concrete Compressive Strength and Parameter Correlation Analysis

Li YiFei, Cao MaoSen, and Magd Abdel Wahab

Abstract The concrete compressive strength (CS) is an important parameter used for durability design and service life prediction of concrete structures in civil engineering projects. It usually has a high nonlinear relationship with the age and main components of concrete, which makes it difficult for traditional regression analysis methods to perform predictive modelling. This study presents a data-driven Kriging model for predicting concrete CS under standard curing period. Two popular machine learning algorithms, namely Artificial Neural Network (ANN) and Support Vector Regression (SVR), are used for comparisons to validate the predictive ability of Kriging model. In addition, a parameter correlation analysis is implemented to reveal the intrinsic association of the selected seven main components of concrete and concrete CS. This study led to the following conclusions: (1) compared with ANN and SVR, the data-driven Kriging model has the highest accuracy in predicting concrete CS, and (2) the results of the parameter correlation analysis coincide with the physical laws of concrete CS.

Keywords Kriging · Concrete compressive strength · Parameter correlation analysis · Data-driven

1 Introduction

Concrete is the most important material in civil engineering and can be regarded as the skeleton of most civil engineering structures. The concrete CS plays a crucial role in the load-bearing properties of concrete and is highly nonlinearly related to the age and several main components of the concrete, which makes its predictive modelling a very difficult task.

L. YiFei (✉) · C. MaoSen
Department of Engineering Mechanics, Hohai University, Nanjing, China
e-mail: li.yifei@ugent.be

L. YiFei · M. Abdel Wahab
Soete Laboratory, Faculty of Engineering and Architecture, Ghent University, Ghent, Belgium

M. Abdel Wahab (ed.), *Proceedings of the 5th International Conference on Numerical Modelling in Engineering*, Lecture Notes in Civil Engineering 311,
https://doi.org/10.1007/978-981-19-8429-7_11

However, considering the production and curing characteristics of concrete, it would be extremely laborious and time-consuming to experimentally measure the concrete CS of multiple specimens. Therefore, if we can develop a concrete CS prediction model with appropriate accuracy based on the moderate experiment data, it will greatly reduce human and financial resources.

Traditional prediction techniques based on regression analysis (linear regression, logistic regression, etc.) often have difficulty modelling with highly non-linear "input–output" datasets [1–3]. In recent years, as machine learning research has been flourishing, some machine learning algorithms have become popular in predictive modelling for their ability to accurately reflect the highly nonlinear mapping relationship between input variables and model outputs, some typical ones such as Kriging, ANN, SVR, etc.

Yeh [4] first explored the possibility of using Artificial Neural Networks (ANNs) to predict the CS of high-performance concrete. The results not only showed that ANN-based strength models were more accurate than those based on regression analysis, but also easily examined the effect of variation in the proportion of each concrete component on concrete CS based on ANN. Guang et al. [5] proposed Multi-layer Feed-forward Neural Networks (MFNNs) to predict the 28-day CS of concrete, and the results demonstrated the practicality and effectiveness of using NNs to predict the concrete strength. Asteris et al. [6] adopted four Conventional Machine Learning (CML) models to train and validate the experimental dataset of concrete CS, and then combined the prediction outputs of these CML models and trained them using ANN to construct a Hybrid Ensemble model (HENSM). The proposed HENSM produced higher prediction accuracy compared to the individual CML models. Oztas et al. [7] utilized experimental data collected in the literature with 187 different HSC concrete mix designs to construct a Neural Network (NN) and predicted the CS and slump of HSC based on this model. Mohammed et al. [8] estimated the CS of concrete containing high volume fly ash (HVFA) based on soft computing techniques, such as ANN and M5P-tree. By comparing with traditional regression methods, the results show that the M5P-tree method has the best performance. Manish A. et al. [9] collected the concrete CS dataset measured by Ultrasonic Pulse Velocity (UPV), and then used multiple regression analysis and artificial neural network to predict concrete CS, respectively, and the results showed that the latter had better prediction results.

The above literature survey shows the wide use of ANN in the prediction of concrete compression strength. In this paper, another popular machine learning algorithm, Kriging model, is used and compared with ANN and SVR. In addition, a parameter correlation analysis was performed to investigate the intrinsic link between several major components of concrete and their compressive strength. In Sect. 2, some brief methodologies on Kriging models and parameter correlation analysis are presented. The three machine learning models are constructed and compared in Sect. 3, then some observed conclusions and detailed results are presented, respectively, in Sect. 4 and Appendix.

2 Theoretical Foundations

2.1 A Methodology of Kriging

As early as 1950's, the Kriging (also known as Gaussian process modelling, GP) was first developed as a spatial interpolation tool in Geostatistics by Krige [10] and proposed formally by Matheron [11] in the 1960s. Sacks et al. [12] introduced Kriging, which was used to represent an "input–output" mapping of an expensive computational model.

Suppose that the model output $Y = \mathcal{M}(x)$ is a realization of a Kriging indexed by $x \in \mathcal{D}_X \subset \mathbb{R}^M$. A Kriging model can be expressed as [13]:

$$Y \approx \mathcal{M}^K(x) = \vartheta^T F(x) + G(x) \tag{1}$$

where ϑ^T represents the transpose of the corresponding regression coefficient vector, $F(x) = [F_1(\boldsymbol{x}), \ldots, F_M(\boldsymbol{x})]$ is the polynomial basis function, and then, $\vartheta^T F(x)$ is the trend of the Kriging model.

In addition, $G(\boldsymbol{x})$ is a Gaussian process with zero mean and covariance function defined as:

$$Cov\big(G(\boldsymbol{x}_i), G(\boldsymbol{x}_j)\big) = \sigma^2 R(\boldsymbol{x}_i, \boldsymbol{x}_j; \boldsymbol{\theta}) \tag{2}$$

where $G(\boldsymbol{x}_i)$ and $G(\boldsymbol{x}_j)$ are, respectively, observation points and new points, σ^2 is the (constant) variance of $G(\boldsymbol{x})$, and $R(\boldsymbol{x}_i, \boldsymbol{x}_j; \boldsymbol{\theta})$ is the correlation function, which describes the "similarity" between $G(\boldsymbol{x}_i)$ and $G(\boldsymbol{x}_j)$ with hyperparameters $\boldsymbol{\theta} = [\boldsymbol{\theta}_1, \ldots, \boldsymbol{\theta}_n]^T$.

The Matérn-5/2 correlation function is selected in this study, and its formulas are given below [14]:

$$R(\boldsymbol{x}_i, \boldsymbol{x}_j; \boldsymbol{\theta}, \nu = 5/2) = (1 + \sqrt{5}\frac{|\boldsymbol{x}_i - \boldsymbol{x}_j|}{\theta} + \frac{5}{3}\left(\frac{|\boldsymbol{x}_i - \boldsymbol{x}_j|}{\theta}\right)^2)\exp\left[-\sqrt{5}\frac{|\boldsymbol{x}_i - \boldsymbol{x}_j|}{\theta}\right] \tag{3}$$

Let us consider $\mathcal{Y} = \{\mathcal{M}(x^{(1)}), \ldots, \mathcal{M}(x^{(N)})\}^T$ is assumed to follow a multivariate Gaussian distribution. By maximizing the likelihood function to estimate the unknown Kriging parameters $\gamma = (\vartheta, \sigma^2, \boldsymbol{\theta})$ as below:

$$\mathcal{L}(\gamma; \mathcal{Y}) = \frac{(detC)^{-1/2}}{(2\pi)^{N/2}} exp[-\frac{1}{2}(\mathcal{Y} - P\vartheta)^T C^{-1}(\mathcal{Y} - P\vartheta)] \tag{4}$$

where the covariance matrix $C = \sigma^2 \boldsymbol{R} + \Sigma_n$, among Σ_n is the noisy responses, $P = [p(x_1), \ldots p(x_N)]^T$ is the N × M regression matrix with element $P_{ij} = p_j(x_i)$.

The partial derivative of Eq. (4) with respect to ϑ and σ^2 to zeros, and thus solving for the hyperparameters $\boldsymbol{\theta}$ can be transformed to solve the following optimization problem:

$$\hat{\theta} = \underset{\theta \in \mathcal{D}_\theta}{argmin}[-log\mathcal{L}(\theta; \mathcal{Y})] \tag{5}$$

Covariance matrix adaptation–evolution strategy (CMA-ES) is a derandomized stochastic search algorithm introduced by Hansen [15], which is adopted to solve the optimization problem later on in Eq. (11). Hence, the optimization problem can be written as:

$$\hat{\theta} = \underset{\theta \in \mathcal{D}_\theta}{argmin}\frac{1}{2}[log(detR) + Nlog(2\pi\sigma^2) + N] \tag{6}$$

2.2 Parameter Correlation Analysis

Parameter correlation analysis provides the most intuitive measure of the sensitivity of the model output Y to the components of the input vector X [16]. Consider a sampling of the input vector $X = \{x(1), x(2), \ldots, x(N)\}$ and the corresponding model output $Y = \{y(1), y(2), \ldots, y(N)\}$, the linear correlation coefficient ρ_i between the ith input and the output is defined as:

$$\rho_i \overset{\text{def}}{=} \rho(X_i, Y) = \frac{\mathbb{E}[(X_i - \mathbb{E}(X_i))(Y - \mathbb{E}(Y))]}{\sigma_i \sigma_Y} \tag{7}$$

where σ_i and σ_Y are the corresponding standard deviations.

However, linear correlation is inaccurate in the presence of strong nonlinear dependence between variables. A more stable estimation method, Spearman's rank correlation, is used in this study, which relies on the monotonicity of the dependence of the two variables rather than linearity [17].

To calculate Spearman correlation coefficient, ρ_S, each component of the input sample and its response are transformed into their rank-equivalents:

$$R_i = \{r_i^{(j)} \in \{1, \ldots, N\} : r_i^{(j)} > r_i^{(k)} \iff x_i^{(j)} > x_i^{(k)} \forall j, k \in \{1, \ldots, N\}\} \tag{8}$$

The rank-transformed model output reads $\boldsymbol{R}_Y = \{r_Y^{(1)}, r_Y^{(2)}, \ldots, r_Y^{(N)}\}$, and then the ρ_S is defined as the linear correlation coefficients between the ranks:

$$\rho_i^S \rho_S(\boldsymbol{X}_i, \boldsymbol{Y}) = \rho(\boldsymbol{R}_i, \boldsymbol{R}_Y). \tag{9}$$

3 Data-Driven Kriging Modelling for Predicting Concrete CS

3.1 Data Preparation

The dataset, 1030 rows and 9 columns, has been sourced from the Machine Learning Repository of University of California, Irvine [4], which is suitable for exploring the predictive performance of supervised and unsupervised learning techniques. The actual concrete CS (MPa) for a given mixture under a specific age (days) was determined from laboratory. In this dataset, eight mixtures that have a significant impact on the concrete CS are considered, as shown in Table 1.

As the initial data set is in raw form (not scaled), it is difficult to model it as a whole to indicate a specific physical behavior. To clarify the significance of the study; prediction of the CS of concrete at standard curing period is performed. Therefore, 120 data sets were selected at a specific age (28 days) as the selected dataset for subsequent modelling in this study.

3.2 Construction and Evaluation of Data-Driven Kriging Model

Considering age of testing (X8) as a constant factor, the selected dataset is a matrix with 120 rows and 8 columns, where main components (X1-X7) are the inputs and concrete CS is the output. Among them, the first 100 sets were used as the design of experiment (DOE) dataset (X, Y) to construct the Kriging model, and the last 20 sets were used as the validation dataset (Xval, Yval) to predict the concrete CS.

Coefficient of determination (R^2), as the most used validation metric for predictive model, is applied to evaluate the accuracy of Data-Driven Kriging Model in this study.

Table 1 The main components affecting the concrete CS

Component	Symbol	Unit
Cement	X1	kg/m^3
Fly ash	X2	kg/m^3
Blast furnace slag	X3	kg/m^3
Water	X4	kg/m^3
Superplasticizer	X5	kg/m^3
Coarse aggregate	X6	kg/m^3
Fine aggregate	X7	kg/m^3
Age of testing	X8	Days

$$R^2 = 1 - \frac{\sum_{k=1}^{N}(Y_k - \tilde{Y}_k)^2}{\sum_{k=1}^{N}(Y_k - \overline{Y_k})^2} \tag{10}$$

Mean absolute percentage error (MAPE) is often preferred for comparing the predicted accuracy of algorithms due to its intuitive percentage representation, which is expressed as follows:

$$\frac{100}{M}\sum_{k=1}^{M}\frac{\left|Y_k - \tilde{Y}_k\right|}{Y_k} \tag{11}$$

where Y_k and $\tilde{Y}_k$ are, respectively, the output value and Kriging predicted value corresponding to the kth set of inputs, $\overline{Y_k}$ is the mean value of the output, and N and M are, respectively, the sample sizes for the DOE and validation datasets.

3.3 Comparison with ANN and SVR

In this section, in order to explore the performance of Data-Driven Kriging Model in predicting the concrete CS, two of the most widely known machine learning algorithms: ANN and SVR are used. The same validation metrics (R^2 and MAPE) are used to evaluate their accuracy. The predicted value of concrete CS based on three methods are, respectively, denoted as $Y_{Kriging}$, Y_{ANN}, Y_{SVR}, as shown in Table 2.

First and most importantly, any machine learning method used for prediction needs to be validated for its own accuracy first, otherwise, the best prediction results obtained are meaningless. Hence, a comparison of the accuracy for the three methods constructed on the same DOE dataset is shown in Fig. 1, from which we can draw some conclusions as follows:

- Intuitively, these predicted strength values based on Kriging model essentially fit the strength observed in Lab. This can be demonstrated that Kriging model has an extremely high accuracy. However, the difference in accuracy between SVR and ANN is not easily observed directly.
- From the quantitative analysis of the accuracy of the three methods, the R^2 of Kriging model is equal to 1, which indicates that the model achieves perfect prediction in the DOE dataset. SVR performs moderately well, while the ANN modelling based on this dataset is less than ideal.

Based on the above analysis, we confirm that the Kriging model can definitely be used to predict the concrete compression strength. In addition, despite the unsatisfactory accuracy of the other two methods, they are still used to predict in order to highlight the outstanding performance of Kriging model. Figure 2 summarizes the

Table 2 Comparison of the prediction accuracy for three models, [MPa]

Y_{Val}	$Y_{kriging}$	Y_{ANN}	Y_{SVR}
30.88	37.55	33.09	33.31
15.34	26.11	14.10	18.03
24.34	31.68	25.85	33.15
23.89	33.73	35.47	31.40
22.93	24.63	23.91	24.55
29.41	31.40	27.26	32.60
28.63	28.55	29.83	29.02
36.80	36.80	33.57	36.32
18.29	18.42	24.62	24.32
32.72	32.67	30.53	32.33
31.42	30.23	30.50	31.33
28.94	30.74	23.90	27.11
40.93	44.32	48.94	50.91
12.18	12.29	10.98	12.51
25.56	25.48	28.45	25.88
36.44	35.67	27.32	21.39
32.96	32.76	29.98	32.49
23.84	23.63	28.31	31.11
26.23	26.14	29.87	27.41
17.96	17.92	13.75	17.53
MAPE	10.106%	14.350%	13.141%

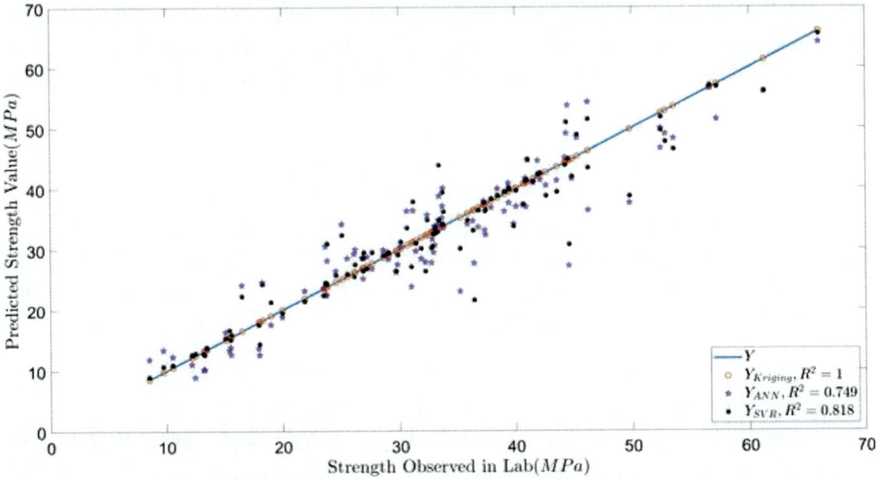

Fig. 1 Comparison on the accuracy of the three models

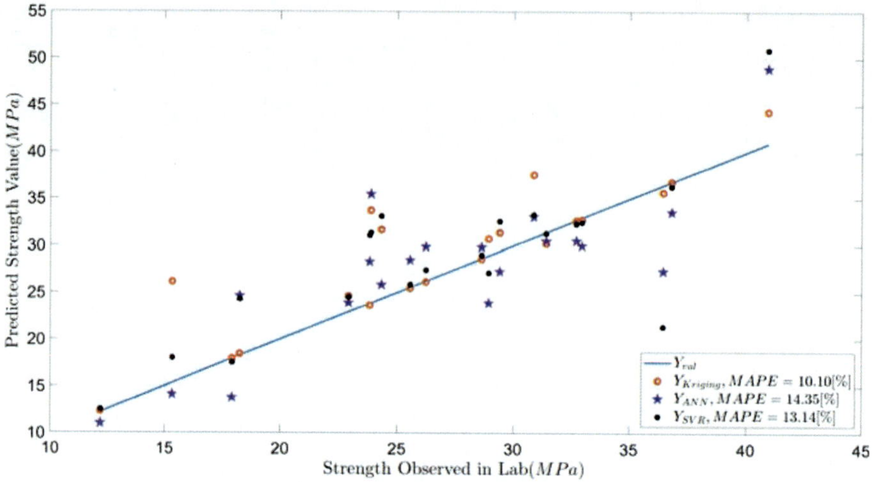

Fig. 2 Comparison on the prediction accuracy of the three models

comparison on the prediction accuracy of the three methods. Some observations are given as follows:

- Different from the accuracy comparison shown in Fig. 1, the Kriging model does not have an overwhelming advantage in terms of prediction accuracy. Although some of the strength values are predicted accurately, others still deviate significantly from the experimental strength values used for validation.
- A quantitative analysis on the prediction accuracy of the three methods based on MAPE, which shows that the Kriging model still has the best accuracy, although the superiority is not obvious. However, it is interesting to notice that even the ANN model with the worst accuracy is not significantly worse than the Kriging model in terms of prediction accuracy.

Although the prediction accuracy of the Kriging model (MAPE $= 10.1\%$) is not ideal, it is sufficient to meet the error control requirements in the field of strength and life prediction. To further insight the intrinsic link between main components and concrete CS, we performed a parameter correlation analysis based on the Kriging model for this purpose. The results of which are shown in Fig. 3, from which some interesting observations can be drawn:

- Of all seven main components, only cement (X1) and fly ash (X2) were positively correlated with the concrete CS, with X1 having the greater correlation.
- Blast furnace slag (X3) shows the greatest negative correlation with concrete CS, and that is well understood because the more blast furnace slag is added, the more holes there are in the concrete at the microscopic level, which is reflected in the reduction of CS at the macroscopic level.

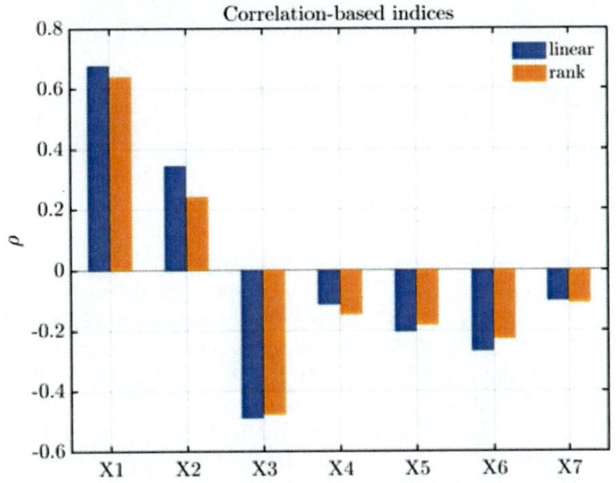

Fig. 3 Correlation analysis of main components and concrete CS

- All main components that show positive and negative correlations with concrete CS can be explained by physical laws.

4 Conclusion

The concrete CS is closely related to a mixture of main components and the links between them show a high non-linear and correlation. Thus, making accurate prediction of concrete CS using classical statistical methods is very difficult.

In this study, a 'black box' modelling based on data-driven kriging approach was conducted for the concrete CS and seven other main components at selected age of testing (28 days). The constructed Kriging model was compared with two other popular machine learning approaches, namely ANN and SVR. Some results are summarized below:

- The constructed Kriging model showed great advantages not only in validating the accuracy of the model itself, but also in presenting the best performance when used to predict concrete CS.
- There is no strict positive correlation between the accuracy of the model itself and the accuracy of the predictions, but models are used for prediction on the assumption that they should be sufficiently accurate to characterize the object of study.

In addition, the results of the parameter correlation analysis provide further insight into the intrinsic association of the seven main components with concrete CS, and all the correlations in the "input–output" are well explained by the laws of physics.

This study provides a good method for concrete CS prediction, but since the discussion is limited to concrete CS prediction at a specific age of testing (28 days), it remains to be explored whether the method is generalizable. Future work will focus on expanding the discussion to investigate the generalizability of the method, and parameter sensitivity analyses will be considered to provide a more comprehensive insight into the intrinsic relationship between main components and concrete CS.

References

1. Ncibi MC (2008) Applicability of some statistical tools to predict optimum adsorption isotherm after linear and non-linear regression analysis[J]. J Hazard Mater 153(1–2):207–212
2. LaValley MP (2008) Logistic regression[J]. Circulation 117(18):2395–2399
3. Su X, Yan X, Tsai CL (2012) Linear regression[J]. Wiley Interdiscip Rev: Comput Stat 4(3):275–294
4. Yeh IC (1998) Modeling of strength of high-performance concrete using artificial neural networks[J]. Cem Concr Res 28(12):1797–1808
5. Ni HG, Wang JZ (2000) Prediction of compressive strength of concrete by neural networks[J]. Cem Concr Res 30(8):1245–1250
6. Asteris PG, Skentou AD, Bardhan A et al (2021) Predicting concrete compressive strength using hybrid ensembling of surrogate machine learning models[J]. Cem Concr Res 145:106449
7. Öztaş A, Pala M, Özbay E et al (2006) Predicting the compressive strength and slump of high strength concrete using neural network[J]. Constr Build Mater 20(9):769–775
8. Kewalramani MA, Gupta R (2006) Concrete compressive strength prediction using ultrasonic pulse velocity through artificial neural networks[J]. Autom Constr 15(3):374–379
9. Mohammed A, Rafiq S, Sihag P et al (2021) Soft computing techniques: systematic multiscale models to predict the compressive strength of HVFA concrete based on mix proportions and curing times[J]. J Build Eng 33:101851
10. Krige DG (1951) A statistical approach to some mine valuation and allied problems on the Witwatersrand. Master's thesis, University of the Witwatersrand, South Africa
11. Matheron G (1963) Principles of geostatistics. Econ Geol 58(2):1246–1266
12. Sacks J, Welch WJ, et al (1989) Design and analysis of computer experiments. Stat Sci 4:409–435
13. Santner T, Williams B, Notz W (2003) The design and analysis of computer experiments. Springer series in Statistics, Springer
14. Lataniotis C, Wicaksono D, et al (2019) UQLab user manual—Kriging (Gaussian process modeling). Report # UQLab-V1.3-105, Chair of Risk, Safety and Uncertainty Quantification, ETH Zurich, Switzerland
15. Hansen N, Ostermeier A (2001) Completely derandomized self-adaptation in evolution strategies. Evol Comput 9(2):159–195
16. Yaghoubi V, Vakilzadeh MK, Abrahamsson TJS (2018) Automated modal parameter estimation using correlation analysis and bootstrap sampling[J]. Mech Syst Signal Process 100:289–310
17. Sedgwick P (2014) Spearman's rank correlation coefficient[J]. Bmj 2014:349.